Columbia University

Contributions to Education

Teachers College Series

No. 534

AMS PRESS
NEW YORK

VOCATIONAL EDUCATION IN AGRICULTURE IN FEDERALLY-AIDED SECONDARY SCHOOLS

A STUDY OF ITS INSTRUCTIONAL AND TRAINING PHASES

By

G. A. SCHMIDT, Ph.D.

Teachers College, Columbia University
Contributions to Education, No. 534

Published with the Approval of
Professor David Snedden, Sponsor

Bureau of Publications
Teachers College, Columbia University
NEW YORK CITY
1932

Library of Congress Cataloging in Publication Data

Schmidt, Gustavus Adolphus, 1880-
Vocational education in agriculture in federally-aided secondary schools.

Reprint of the 1932 ed., issued in series: Teachers College, Columbia University. Contributions to education, no. 534.
Originally presented as the author's thesis, Columbia.
Bibliography: p.
1. Agricultural education--United States.
2. Agriculture--Study and teaching (Secondary)--United States. I. Title. II. Series: Columbia University. Teachers College. Contributions to education, no. 534.
S533.S4 1972 630'.7'1273 75-177802

ISBN 0-404-55534-9

Reprinted by Special Arrangement with Teachers College Press, New York, New York

From the edition of 1932, New York
First AMS edition published in 1972
Manufactured in the United States

AMS PRESS, INC.
NEW YORK, N. Y. 10003

ACKNOWLEDGMENTS

The writer gratefully acknowledges the helpful criticisms, the time, and the courtesies which Professor David Snedden, chairman of his dissertation committee, and Professors Florence B. Stratemeyer and John D. Willard, the other two members of his dissertation committee, have so generously given him.

He also takes this opportunity to express his thanks and appreciation to all the persons who, by filling out and returning to him questionnaires, furnished him the facts upon which this study is based. Without this assistance and response this study would not have been possible.

G. A. S.

CONTENTS

TABLES

VOCATIONAL EDUCATION IN AGRICULTURE IN FEDERALLY-AIDED SECONDARY SCHOOLS

PART I

INTRODUCTION

This study is devoted to an investigation of the instruction and training being offered at the present time to boys enrolled in all-day or full-time federally-aided vocational agriculture classes in the public high schools of the United States. It is in no way concerned with those other forms of vocational education in agriculture of less than college grade which in certain localities are being offered in secondary schools: namely, evening class instruction given to adult farmers; part-time class instruction given to the out-of-school farm youths; day-unit class instruction given to regularly enrolled high school pupils but taught only once or twice a week by an itinerant teacher; instruction and training in agriculture given in special agriculture schools of a secondary nature. Neither is this study concerned with the teaching of general or vocational agriculture in high schools receiving no federal aid for such instruction under the provisions of the Smith-Hughes Act.

Enrollment in Vocational Agriculture Classes in Secondary Schools

According to the Fourteenth (1930) Annual Report of the Federal Board for Vocational Education, 3,905 public secondary schools—practically all of them public high schools—received, during the school year 1929-1930, federal aid, under the provisions of the Vocational Education or Smith-Hughes Act, for the teaching of all-day classes in vocational agriculture. There were enrolled during the school year 1929-1930 in the vocational agriculture classes of these 3,905 secondary schools 113,728 all-day or full-time pupils. These facts are mentioned here merely to show that the teaching of vocational agriculture in public high schools in all-day classes in the United States is a large and important undertaking of considerable economic and social significance, affecting, as it now does annually, the lives of more than 100,000 young people in our secondary schools.

From the beginning the development of the teaching of vocational agriculture in secondary schools under the provisions of the Smith-Hughes Act has been exceedingly rapid, as is shown in Table 1. The work started in 1917. Since then there has been a rather uniform growth both in the number of schools undertaking the work and in the number of pupils enrolled in the vocational agriculture classes in these schools.

TABLE 1

GROWTH IN NUMBER OF FEDERALLY-AIDED SECONDARY SCHOOLS OFFERING INSTRUCTION IN VOCATIONAL AGRICULTURE IN ALL-DAY CLASSES, AND GROWTH OF PUPIL ENROLLMENT IN THESE CLASSES*

School Year Ending June 30th	Number of Schools	Number of Pupils
1917	0	0
1918	609	15,453
1919	863	19,933
1920	1,375	31,301
1921	1,610	40,763
1922	1,935	52,961
1923	2,165	57,978
1924	2,480	65,358
1925	2,778	70,958
1926	3,081	82,431
1927	3,339	89,390
1928	3,553	96,941
1929	3,788	106,111
1930	3,905	113,728

* Taken from the Twelfth and the Fourteenth Annual Reports of the Federal Board for Vocational Education.

The Vocational Education Act

In 1917 Congress passed the Vocational Education Act, or the Smith-Hughes Act,[1] as it is more commonly called. This Act was passed by Congress to promote and to encourage public vocational education of less than college grade throughout the nation in agriculture, trades and industries, and home economics.

The Act promotes and encourages vocational education in these

[1] The Federal Vocational Education Act, Public Document No. 347, 64th Congress, S. 703.

three fields by appropriating annually $3,000,000 for the purpose of coöperating with states in paying the salaries of teachers, supervisors, and directors of agricultural subjects; $3,000,000 for the purpose of coöperating with the states in paying salaries of teachers of trade, home economics, and industrial subjects; and $1,000,000 for the purpose of coöperating with the states in training teachers, supervisors, and directors of vocational agriculture of less than college grade and teachers of vocational trade, home economics, and industrial subjects.

The federal appropriations for the teaching of vocational agriculture and vocational home economics, under the provisions of the Smith-Hughes Act, were substantially increased by Congress in 1929 by the passage of the George-Reed Act.[2] In this Act the additional federal appropriations, supplementing those in the Smith-Hughes Act, for vocational education in agriculture and for vocational education in home economics, will amount to $2,500,000 in 1934; or, in other terms, $1,250,000 for each of the two types of work.

The Fundamental Purpose of the Smith-Hughes Act

The fundamental purpose underlying federal aid for the promotion of vocational education of less than college grade is to give persons an opportunity, in schools and classes under public supervision or control, to fit themselves for useful employment in the home, on the farm, in the trades, and in the industries; and to give those engaged in these four large groups of occupations opportunity to improve themselves in their occupations by attending evening or part-time classes in vocational education.

The purpose underlying the passage of the Smith-Hughes Act is clearly expressed in the Report of the Commission on National Aid to Vocational Education[3] in the following quotation:

> "While many different kinds and grades of vocational education will always be required, the kind most urgently demanded at the present time is that which will prepare *workers* for the more common occupations in which the great mass of our people find useful employment.
>
> "There is a great and crying need of providing vocational education of this character for every part of the United States

[2] An Act to provide for the further development of vocational education in the several states and territories. Public Document No. 702, 70th Congress, S. 1731.

[3] *Report of the Commission on National Aid to Vocational Education*, House Document No. 1004, 63rd Congress, 2nd Session.

—to conserve and develop our resources; to promote a more productive and prosperous agriculture; to prevent the waste of human labor; to supplement apprenticeship; to increase the wage-earning power of our productive workers; to meet the increasing demand for trained workmen; to offset the increased cost of living. Vocational education is therefore needed as a wise business investment for this Nation; because our national prosperity and happiness are at stake and our position in the markets of the world can not otherwise be maintained.

"The social and educational need for vocational training is equally urgent. Widespread vocational training will democratize the education of the country: (1) by recognizing different tastes and abilities and by giving an equal opportunity to all to prepare for their life work; (2) by extending education through part-time and evening instruction to those who are at work in the shop or on the farm."

President Hoover, also, clearly defines the purpose of vocational education of less than college grade as promoted under the provisions of the Smith-Hughes Act. He expresses this purpose as follows:

"The essential purpose of the Smith-Hughes Act is to provide for the needs of our youths who do not enter our higher technical and professional educational institutions. Vocational training for the commoner wage-earning pursuits and skilled trades is equally as essential as is training for the professions. The humblest worker equally with the youth who proposes to enter the professions, has a right to the sort of training he needs for the occupation by which he proposes to earn his livelihood and support his family, and through which he will render his service to the community in getting the community's work done. We cannot in fairness continue to provide specialized education free to the few who propose to enter the professions, while denying education to the many for the commoner vocations.

"There is in fact no better economy than the economy of adequate training for the pursuits of agriculture, commerce, industry, and the home. Our youth must enter into these pursuits, and it is on all counts in the public interest that they be well trained for them. To provide such training is clearly a public responsibility. Education in general, including vocational education for the youth, is Democracy's most important business. Democracy in education means that in the field of education opportunity shall be extended equally to all—to give all a fair start. This is the educational ideal inspiring those who are administering the Federal Vocational Education

Act; it is the ideal which inspired Congress in passing the Act; and it is traditionally the ideal of education in our democracy."[4]

The Fundamental Purpose of Vocational Education in Agriculture of Less Than College Grade

In section ten of the Smith-Hughes Act the purpose of vocational education in agriculture of less than college grade is set forth as follows:

1. "The controlling purpose of such education shall be to fit for useful employment;
2. "that such education shall be of less than college grade; and
3. "be designed to meet the needs of persons over fourteen years of age, who have entered upon or who are preparing to enter upon the work of the farm or of the farm home."

If vocational education in agriculture of less than college grade must fit persons for useful farm employment and must be designed to meet the needs of persons who have entered upon or who are preparing to enter upon the work of the farm, the objective of such education clearly shows itself to be:

> "To train prospective farmers for proficiency in some specific farming occupation, or to improve those already engaged in farming in order that they may become more efficient in the farming occupation in which they are engaged."

The objective quoted above expresses the point of view held by leaders in vocational education[5] in regard to the aims and purposes of vocational education in agriculture of less than college grade. It expresses also the point of view of a Committee on Objectives in Vocational Agriculture, recently called by the Federal Board for Vocational Education.[6]

[4] Taken from a letter written by President Hoover to J. C. Wright, Director of the Federal Board for Vocational Education.

[5] To substantiate this statement reference is made to the following authorities: Bagley, William C. and Keith, John A. H., *An Introduction to Teaching*, p. 268. Bobbitt, Franklin, *How To Make a Curriculum*, p. 66. Eaton, Theodore H., *Education and Vocations*, p. 145. Lee, Edwin A., *Objectives and Problems of Vocational Education*, p. 93. Prosser, Charles A. and Allen, Charles R., *Vocational Education in a Democracy*, p. 8. Snedden, David, *Vocational Education*, p. 8. Wright, J. C. and Allen, Charles R., *Efficiency in Vocational Education*, p. 48.

[6] Report of Committee on Objectives in Vocational Agriculture. Miscellaneous Publications 1046, Federal Board for Vocational Education. Same report is also found in *Training Objectives in Vocational Education in Agriculture*, Bulletin 153, Federal Board for Vocational Education.

General and Vocational Education

Two general schemes are followed in giving preparatory or pre-employment vocational training.

In one of these schemes the vocational training follows general education. The implication here is that a pupil will first acquire all the general school education he wishes or which he has time and opportunity to acquire, and that he will have completed his general education before he enters upon a program of vocational education; and that after he has entered the vocational school, he will give all his school time to his vocational education.

Many authorities in the fields of both vocational and general education favor this plan for preparatory vocational education.

In the other scheme the vocational training parallels that of the general education, and the two are pursued simultaneously, the individual devoting part of the school day to general education and part to vocational education. Many authorities in both general and vocational education favor this latter scheme in certain kinds of vocational education of less than college grade. The particular kinds of vocational education for which they favor this plan are those in which the following conditions prevail:

1. Where the school can give real and effective vocational instruction which will prepare the trainees for competency in the occupation for which the instruction is offered.
2. Where the school can provide, or make provisions for, a real occupational environment for giving to the pupils adequate participating experience or training which functions and is needed in the occupation for which the training is being offered.
3. Where the pupils are old enough to profit by the instruction and training.

At the present time practically all the preparatory vocational education in agriculture of less than college grade given in our public secondary schools under the provisions of the Smith-Hughes Act is that which is offered in conjunction with general secondary education. This point is discussed in detail in Part X of this study.

Vocational Efficiency the Aim of Vocational Education

Whether preparatory, that is, pre-employment, vocational education in agriculture follows general school education, or whether it is given simultaneously with general school education, it must in either case be genuine or real vocational education. There is only one measure of a preparatory vocational training course. That is the degree to which it prepares pupils for competency in the specific occupation for which they are being vocationally trained.

Since boys enrolled in all-day vocational agriculture courses in secondary schools pursue both vocational and general educational subjects, it follows naturally that the vocational agriculture subjects in the vocational agriculture curriculum of our secondary schools, and the training offered by these schools must be such as will make specific contributions to competency in the farming occupation for which the instruction and training is being offered. Also, the general educational subjects offered in the vocational agriculture curriculum must be such as will make a specific contribution to the attainment of the aims of general education—the development of the civic, social, cultural, and physical well-being of the pupils. To ascertain how these two aims are being attained at the present time in the high schools of the United States is the purpose underlying this study.

PART II

THE PROBLEM, AND OUTLINE OF THE PROCEDURE FOLLOWED IN SOLVING IT

Vocational education in agriculture—that is, the strictly vocational instruction and training—as promoted under the provisions of the Smith-Hughes Act has a very specific function: that of fitting persons for useful farm employment. This particular function is clearly defined in the Act and has been explained and illustrated by means of quotations in Part I of this study.

In addition the Smith-Hughes Act also clearly specifies that any general education which may be necessary to build a well-rounded course of training may accompany the strictly vocational instruction and training, but no federal funds appropriated in the Act can be used for this purpose. The following quotation from Section 9 of the Act describes this feature:

> "The cost of instruction supplementary to the instruction in agriculture and in trade, home economics, and industrial subjects provided for in this Act, necessary to build a well-founded course of training, shall be borne by the State or local communities, and no part of the cost thereof shall be borne out of the appropriations herein made."

This study proposes to make an investigation of various important aspects of the vocational instruction and training now being offered to all-day or full-time pupils enrolled in federally-aided agriculture classes in our public high schools and to make, also, an investigation of the general education offered to the boys enrolled in these classes. The purposes of the study are to gather pertinent functioning facts relative to both phases of education, that is, the vocational and the general, which are being offered to these boys, and to discover possible ways of improving the training in each kind of education.

The Problem

The general problem involved in this study divides itself into two main divisions or problems: (1) To what degree does the instruction and training offered in all-day vocational agriculture classes in our secondary schools seem to conform to factors characterizing an effective vocational training program? and (2) To what degree are the required general educational subjects in the vocational agriculture curriculum such as offer the best possible instruction for effective citizenship or for making a well-rounded course of training?

In solving these two main problems it seemed necessary to make a study of the following sub-problems:

1. What factors characterize effective vocational instruction and training for boys enrolled in all-day vocational agriculture courses in secondary schools?

2. What general educational (non-vocational) subjects are of most worth to boys taking the all-day vocational agriculture course in secondary schools and should be made required subjects (constants) for them?

3. What are the requirements in the all-day vocational agriculture schools and classes as formulated in the State Plans for Vocational Education of the various states?

4. What types of boys are enrolled in all-day vocational agriculture classes in secondary schools?

5. What is the general nature of the all-day vocational agriculture course in secondary schools?

6. What procedures and methods are used in the all-day vocational agriculture classes in secondary schools in training boys for farming?

7. What is the nature of the general education (non-vocational instruction) offered boys enrolled in all-day vocational agriculture courses in secondary schools?

8. How is the all-day vocational agriculture course in secondary schools administered?

Parts III to X of this study are given over to the solution, respectively, of the above-mentioned eight sub-problems.

The Procedure Followed in Gathering Functioning Data on the Sub-Problems

To secure expert opinion on factors characterizing an effective vocational training program in agriculture for boys enrolled in all-day vocational agriculture courses in our secondary schools (sub-problem 1) the questionnaire on page 10 was sent to sixty persons considered competent to pass judgment on such a problem. The names of these sixty persons were suggested by two men, one an outstanding authority in all phases of vocational education of less than college grade, and the other a man prominent in vocational education in agriculture. The list of names to whom this questionnaire was sent included the following:*

State directors of vocational education	14
State supervisors of vocational agriculture	10
Teacher trainers in vocational agriculture	13
State supervisors of trade and industrial education	6
Members of the staff of the Federal Board for Vocational Education, five representing agriculture, three representing the trades and industries, and three representing home economics	11
Persons from various other fields well qualified to give expert opinion on the problem	6

The thirteen factors characterizing an effective vocational training program in agriculture for boys enrolled in vocational agriculture courses in our secondary schools, upon which expert opinion was sought, were derived from study of the available literature bearing on the problem.[1] The writer selected and reworded from a list of sixteen factors frequently found in books and bulletins discussing efficiency in vocational education, all those which he thought were applicable to the particular field of vocational education under discussion.

QUESTIONNAIRE TO EXPERTS IN THE FIELD OF VOCATIONAL EDUCATION

In connection with a study I have undertaken, I am gathering expert opinion on factors characterizing the instruction and training in effective *vocational education in agriculture* of less than college grade for all-day or full-time pupils enrolled in vocational agriculture courses in our secondary schools.

* The names of the persons who returned the questionnaire are given in the Appendix.

[1] Lee, Edwin A., *Objectives and Problems in Vocational Education.*

Mays, Arthur B., *An Introduction to Vocational Education.*

National Committee of the A. V. A. on Advanced Courses in Vocational Education, *Training Administrators of Vocational Education,* Bulletin 141, Federal Board for Vocational Education.

Prosser, Charles A. and Allen, Charles R., *Vocational Education in a Democracy.*

Prosser, Charles A. and Allen, Charles R., *Have We Kept the Faith?*

Schmidt, G. A., *Efficiency in Vocational Education in Agriculture.*

Snedden, David, *Vocational Education.*

Wright, J. C. and Allen, Charles R., *Efficiency in Education.*

Wright, J. C. and Allen, Charles R., *Efficiency in Vocational Education.*

Wright, J. C. and Allen, Charles R., *Supervision of Vocational Education.*

I shall greatly appreciate your courtesy if you will check those factors enumerated below which you think are characteristic of effective vocational instruction and training in agriculture of less than college grade for all-day pupils. Please use blank spaces for mention of any other factors which you think should be added to the list. Replies will be strictly confidential matters and the results of this study will show only cumulative results.

_____ 1. The aim of the all-day vocational agriculture course offered in our secondary schools is to train prospective farmers for proficiency in the farming occupation in which they are fitting themselves to engage.

_____ 2. The instruction and training in vocational agriculture is given to a selected group of pupils who need it, want it, and are able to profit by it.

_____ 3. The vocational training environment is the working environment itself, or a replica of the working environment.

_____ 4. The teaching content (subject matter) of the all-day vocational agriculture course is such as specifically functions in the farming occupation for which the pupils are being vocationally trained.

_____ 5. The supervised training experiences (projects and supplementary farm practice work) in which the pupils enrolled in the all-day vocational agriculture classes engage are such as specifically function in the farming occupation for which the pupils are being vocationally trained.

_____ 6. The basis of the teaching content of the vocational agriculture courses offered in secondary schools is obtained through the process of job analysis from masters in the occupation, and not from theorists.

_____ 7. The vocational instruction and training is given on farm enterprises and farm jobs occurring in the farming occupation for which the training is offered, and not on exercises or pseudo jobs.

_____ 8. The important manipulative habits and thinking habits required for competency in the farming occupation for which the training is being offered, are developed in the pupils taking the instruction and training.

_____ 9. The training jobs are carried on in accord with the best and most economical occupational standards.

_____ 10. The vocational training helps the pupils to capitalize their interests, aptitudes, and abilities to the highest degree possible.

_____ 11. The instructional needs of the pupils are met at the time when those pupils most require help, and individual instruction is given whenever necessary to the progress of any pupil.

_____ 12. The vocational instruction and training is continued until the trainee possesses a degree of vocational competency sufficient to enable him to secure and hold employment in the occupation in which he is fitting himself to engage.

_____ 13. The administration of the all-day vocational agriculture course is elastic, and not rigid or highly standardized.

_____ 14. ____

_____ 15. ____

Name of person reporting ______________________________

Position ____________________________ State __________

The information gathered on the questionnaire seeking expert opinion on factors characterizing an effective training program in all-day vocational agriculture classes in secondary schools is recorded in Tables 2 and 3 on pages 16, 17 and 18. It is made the basis of discussion in Part III of this study.

To secure expert opinion in regard to the general educational subjects which are of most worth to boys enrolled in vocational agriculture courses, and which should be constants in the vocational agriculture curriculum (sub-problem 2), the questionnaire on page 13 was sent to sixty well-known educators. The list of names was selected by two competent, disinterested educators. The list included the members of The Committee on Rural Education of the National Society for the Study of Education which prepared Part I of the society's *Thirtieth Yearbook*, entitled "The Status of Rural Education," the members of the National Education Association Committee on the Reorganization of Secondary Education, and a number of other noted educators in the field of general education, making a total of sixty in all to whom the questionnaire on page 13 was sent.[2] The tabulated returns on this questionnaire are found on page 24 of this study.

To gather data necessary to ascertain the requirements in vocational education in agriculture in all-day schools and classes as formulated in State Plans for Vocational Education (sub-problem 3), the writer asked each of the forty-eight state supervisors of vocational agriculture for a copy of the State Plans for Vocational Education of his state. Twenty-seven copies were received. These were analyzed and studied, and are discussed in Part V of this study.

Data essential to the solution of sub-problem 4, types of boys enrolled in all-day vocational agriculture courses in secondary schools, and of sub-problem 5, the general nature of the all-day vocational agriculture courses in secondary schools, were obtained from a questionnaire sent to one hundred and twenty-five teachers of vocational agriculture in the same high schools from which the state supervisors of vocational agriculture had secured data essential to the solution of sub-problem 6. The data relative to the solution of sub-problems 4 and 5 are found and discussed in Parts VI and VII of this study.

[2] The names of the persons who returned this questionnaire are given in the Appendix.

QUESTIONNAIRE TO EXPERTS IN THE FIELD OF SECONDARY EDUCATION

In connection with a study I have undertaken, I am trying to ascertain expert opinion as to the *kinds of general educational subjects* of most worth to boys taking the vocational agriculture courses in our secondary schools. I am assuming the following:

1. That boys enrolled in vocational agriculture courses in our secondary schools are taking the work in vocational agriculture as preparation for farming.
2. That the vocational agriculture courses are offered in high schools located in agricultural communities.
3. That the vocational agriculture subjects in the vocational agriculture curriculum in the secondary school are such as will make specific contributions to vocational competency in the farming occupation for which the vocational training is being offered.
4. That the general educational subjects in the vocational agriculture curriculum must be such as will make specific contributions to the attainment of the general educational aims (except the vocational aim) of the secondary school.

Will you please check below the *general educational subjects* (as they are now commonly being taught) which you think should be *required subjects* for boys taking the vocational agriculture course in secondary schools? Replies will be confidential matters. Only cumulative results will be mentioned in this study.

_____ English
_____ Latin
_____ Spanish
_____ French
_____ German
_____ Algebra
_____ Geometry
_____ General mathematics
_____ Practical arithmetic
_____ Ancient history
_____ European history
_____ American history
_____ World history
_____ Civics
_____ Social problems or sociology
_____ Economics
_____ General science
_____ Biology
_____ Physics
_____ Chemistry
_____ Psychology
_____ Physiology
_____ Physical and health education
_____ Typewriting
_____
_____

Name of person reporting _______________ Institution _________

In order to gather data from which could be deduced facts relative to the procedures and methods used in the all-day vocational agriculture classes in secondary schools in training boys for farming (sub-problem 6), five copies of a questionnaire were sent to each of the forty-eight state supervisors of vocational agriculture, these supervisors being asked to gather the information called for on the questionnaires from the teachers of vocational agriculture in the first five departments of vocational agriculture in high schools which they would visit on their next trip in the field. Thirty-four supervisors returned one hundred and fifty-three questionnaires, making a total of one hundred and fifty-three departments of vocational agriculture which were studied. The classified information assembled on the questionnaire is found in Part VIII of this study.

Information relative to the nature of the general educational (non-vocational) subjects pursued by boys in all-day vocational agriculture classes in secondary schools (sub-problem 7) and information relative to the administration of the vocational courses in secondary schools (sub-problem 8) were also secured from teachers of vocational agriculture on the same questionnaire which was used to secure data for the solution of sub-problems 4 and 5. The data collected on sub-problems 7 and 8, together with a discussion of these two problems, are found in Parts IX and X of this study.

PART III

FACTORS CHARACTERIZING AN EFFICIENT VOCATIONAL TRAINING PROGRAM IN AGRICULTURE IN ALL-DAY CLASSES IN SECONDARY SCHOOLS

The fact was mentioned on page 6 that vocational education in agriculture of less than college grade is education for vocational efficiency in farming, and that the vocational instruction and training offered in all-day vocational agriculture courses in our secondary schools must be such as will make specific contributions to competency in the farming occupation for which the instruction and training is offered.

What factors characterize an efficient preparatory or pre-employment vocational training program in agriculture in all-day classes in secondary schools is the problem made the basis of discussion in this part of the study. As was mentioned on page 10, the writer, as a result of a study of the literature bearing upon efficiency in vocational education, selected thirteen factors characterizing an effective pre-employment or preparatory training program in vocational education in agriculture. In order to secure expert opinion upon these thirteen factors the questionnaire on page 10, stating these thirteen factors, was sent to sixty persons thought competent to pass expert judgment on such a problem. The selection of these persons was described on page 10. As stated on the questionnaire, the persons to whom the questionnaire was sent were asked to check those factors which they thought to be characteristic of an effective preparatory training program in vocational education in agriculture of less than college grade, and to use the blank spaces on the questionnaire for the mention of any other factors which they thought should be added to the list.

The Response to the Questionnaire

The replies received on this questionnaire are shown in Table 2. An inspection of the data in this table shows that

TABLE 2

DISTRIBUTION OF FIFTY-ONE OPINIONS ON EACH OF THE THIRTEEN FACTORS CHARACTERIZING AN EFFECTIVE TRAINING PROGRAM IN VOCATIONAL EDUCATION IN AGRICULTURE IN ALL-DAY CLASSES IN SECONDARY SCHOOLS

Factor	Number Checked	Number Slightly Modified	Number Not Checked	Per Cent Checked
1. The aim of the all-day vocational agriculture course offered in our secondary schools is to train prospective farmers for proficiency in the farming occupation in which they are fitting themselves to engage	46	3	2	90.2
2. The instruction and training in vocational agriculture is given to a selected group of pupils who need it, want it, and are able to profit by it ..	47	3	1	92.1
3. The vocational training environment is the working environment itself, or a replica of the working environment	46	3	2	90.2
4. The teaching content (subject matter) of the all-day vocational agriculture course is such as specifically functions in the farming occupation for which the pupils are being vocationally trained	47	3	1	92.1
5. The supervised training experiences (projects and supplementary farm practice work) in which the pupils enrolled in the all-day vocational agriculture classes engage are such as specifically function in the farming occupation for which the pupils are being vocationally trained	47	1	3	92.1
6. The basis of the teaching content of the vocational agriculture courses offered in secondary schools is obtained through the process of job analysis from masters in the occupation, and not from theorists	42	4	5	82.4
7. The vocational instruction and training is given on farm enterprises and farm jobs occurring in the farm-				

TABLE 2 (*Continued*)

Factor	Number Checked	Number Slightly Modified	Number Not Checked	Per Cent Checked
ing occupation for which the training is offered, and not on exercises or pseudo jobs	45	2	4	88.2
8. The important manipulative habits and thinking habits required for competency in the farming occupation, for which the training is being offered, are developed in the pupils taking the instruction and training	43	3	5	85.5
9. The training jobs are carried on in accord with the best and most economical occupational standards	45	2	4	88.2
10. The vocational training helps the pupils to capitalize their interests, aptitudes, and abilities to the highest degree possible	42	4	5	82.4
11. The instructional needs of the pupils are met at the time when those pupils most require help, and individual instruction is given whenever necessary to the progress of any pupil	46	1	4	90.2
12. The vocational instruction and training is continued until the trainee possesses a degree of vocational competency sufficient to enable him to secure and hold employment in the occupation for which he is fitting himself to engage	39	0	12	76.5
13. The administration of the all-day vocational agriculture course is elastic, and not rigid or highly standardized	45	1	5	88.2

more than seventy-six and one-half per cent or more of the fifty-one persons who responded to the questionnaire checked each of the thirteen factors, thus indicating that they regarded these factors as characteristic of an effective preparatory training program in vocational agriculture.

A vocational training program, regardless of where and how it is offered or administered, must be judged from the occupational point of view and not from the academic schoolroom point of view. The thirteen factors enumerated on the questionnaire on page 10 may characterize conditions far in advance of much of the actual present-day practice in secondary schools; nevertheless they characterize conditions which must be regarded as essential to effective preparatory vocational instruction and training and which, when any study of what is actually being done in all-day vocational agriculture courses in secondary schools is being made, must be regarded as representing effective conditions for efficient vocational instruction and training.

Additional Factors Inserted on the Questionnaires

Numerous additional factors had been inserted or recorded by recipients on the returned questionnaires. Fifteen of the more important of these are given in Table 3. These were submitted individually by the persons returning the questionnaires. Many of these additional factors might perhaps have been checked by others who returned the questionnaire had these factors been originally submitted for checking.

TABLE 3

ADDITIONAL FACTORS INSERTED ON QUESTIONNAIRE SEEKING EXPERT OPINION ON FACTORS CHARACTERIZING AN EFFECTIVE TRAINING PROGRAM IN VOCATIONAL EDUCATION IN AGRICULTURE IN ALL-DAY CLASSES IN SECONDARY SCHOOLS

1. The instruction is given by teachers both occupationally and professionally trained.
2. Pupils are assisted in placement or establishment in the occupation for which they are being trained.
3. Efficient use is made of educational procedures, methods, and devices with regard to: (1) character of the teaching job, (2) characteristics of the instructional group, and (3) working conditions.
4. There is a close correlation of the supervised farm practice work with both the science and the agricultural subject matter.
5. The vocational instruction should develop a pride of craftsmanship fitting the vocation.
6. The traits and responsibilities concomitant to a farmer's career are developed as a part of the vocational training.
7. Sufficient practice in applying the content taught under "close supervision" must be provided to assure the formation of proper working habits.
8. Interests, ideals, attitudes, and appreciations receive recognition appropriate to each enterprise, job, and problem arising.
9. Appreciation of farm life and living conditions encountered in the vocation are taught as an incidental part of the strictly vocational work.
10. Interests and attitudes toward farming as a satisfying way of living and service are developed.

Table 3 (*Continued*)

11. The instruction should be on an individual basis permitting the learner to progress according to his ability to master the subject.
12. Coöperation should be developed to the extent that it will carry over into the marketing of agricultural commodities to the mutual interest of all concerned.
13. The pupils are trained in business principles and practices.
14. The project and the practice teachers are on duty throughout the producing season, with the exception of ten days' vacation in summer.
15. Day vocational instruction must make available wide opportunity for tie-up in natural setting, use of class and group projects, to supplement the more isolated individual projects and farm practice skills.

Comments Made on the Questionnaires

Many of the questionnaires returned by persons who were asked to express their opinions on factors characterizing an effective training program in vocational agriculture for boys enrolled in all-day classes in secondary schools contained some comments which are worthy of mention here because they help to clarify some situations. An enumeration of the more important of these comments, together with a few remarks inserted after each by the writer of this study, follows:

> 1. "Instructors are still teaching courses or subjects, and supervised farming remains detached. Teacher trainers and supervisors are at fault, plus the traditional school system in which the agriculture teacher finds himself."

This is undoubtedly true in many of the situations with which the writer is familiar. As mentioned on page 40, the lack of emphasis regarding this matter in the State Plans for Vocational Education is another factor which may be largely responsible for this situation.

> 2. "I would prefer to teach what a boy is doing or planning to do, rather than have a boy do what he is taught."

If the teaching content (subject matter) of the all-day vocational agriculture course is such as specifically functions in the farming occupation for which the pupils are being vocationally trained (Factor 4); and if the supervised training experiences (projects and supplementary farm practice work) in which the pupils enrolled in vocational agriculture classes engage are such as specifically function in the farming occupation for which the pupils are being vocationally trained (Factor 5), then undoubtedly the instructor is teaching what the boys are doing and planning to do. Basing the teaching on pupil enterprises, functioning in the occupation in which the pupils are fitting themselves to engage, is ideal vocational teaching.

3. "Masters in the occupation should be interpreted as good, successful farmers in the community."

This is the common meaning assigned to the term "masters in the occupation" when used in connection with vocational training in agriculture.

4. "I judge that these thirteen factors are those which should apply in vocational schools and not necessarily those which are now found in such schools."

Truly, the factors characterize an efficient and effective training program toward the realization of which the vocational training should be directed. This was the verdict of the experts in vocational education that were consulted about this problem.

5. "We still have a long way to go if we are to have real vocational work (in agriculture in secondary schools). I like the characteristics, but we are a long way from some of them."

The work in vocational agriculture in secondary schools is comparatively new. First thoughts and efforts were directed largely toward getting it started and going. We have now reached a state where considerable attention can very profitably be given to making it as effective as is possible. This study is undertaken as a step in this direction.

6. "Many of these factors are stressed or attempted but not always with success."

To improve any situation requires constant effort. A continuous effort, if whole-heartedly undertaken, will eventually bring some beneficial results. As a rule, it takes a long time to change traditional school practices. The history of education shows, however, that these practices have changed considerably and will continue to change as constant efforts for improvement are made.

7. "Proper guidance in the junior high school is essential in getting a highly selected group in vocational agriculture courses."

This statement is true, but adequate publicity and personal contact with prospective pupils and the parents of these pupils by the teacher of vocational agriculture are perhaps even more important in securing the kind of pupils for whom the work is intended.

> 8. "Supplementary interpretative science should be included in the subject matter taught in the vocational agriculture classes."

Although not so mentioned in Factors 4, 6, 7, and 8, to teach such supplementary interpretative science in close connection with farm jobs is a commonly recommended practice, quite generally followed. This practice is referred to on page 19 in one of the factors added to the original thirteen.

> 9. "Assuredly some of the conditions implied in the thirteen factors will be attained only with the greatest difficulty and probably not for some considerable time; but to a complete and adequate program in vocational education in agriculture, I cannot see where any one of these points could be omitted."

An ideal rarely, if ever, reaches a 100-per-cent attainment. To approximate an ideal in any large movement always involves overcoming great difficulties and always takes much time. Successful attainments in any line of human endeavor do not "just happen." They are the result of hard effort extended over much time.

Summary and Conclusions

Seventy-six and one-half per cent or more of the fifty-one persons returning the questionnaire seeking expert opinion on factors characterizing an effective training program in all-day vocational agriculture classes in secondary schools regarded each of the thirteen factors listed in Table 2 on pages 16-17 as characteristic of such an effective program. These factors, together with the more important ones of those added to the list, as given in Table 3 on page 18, may characterize conditions far in advance of those quite generally found at the present time in all-day vocational agriculture classes in secondary schools; nevertheless they represent conditions which seem essential in getting efficient results in a preparatory vocational training program in agriculture. These thirteen factors will therefore be used in later parts of this study in an attempt to compare existing conditions and practices found and followed in all-day vocational agriculture classes in secondary schools; and to make suggestions for the future development of the work which will enable it to render a greater social service.

PART IV

GENERAL EDUCATIONAL SUBJECTS OF MOST WORTH TO BOYS ENROLLED IN ALL-DAY VOCATIONAL AGRICULTURE COURSES IN SECONDARY SCHOOLS

It is a generally accepted fact that the vocational agriculture subjects in the vocational agriculture curriculum in the secondary school should be such as will make specific contributions to vocational competency in the farming occupation for which the vocational training is being offered, and that the general educational subjects in the vocational agriculture curriculum should be such as will make specific contributions to the attainment of the general educational aims of secondary education, these aims being the civic, cultural, moral, and health growths of the pupils.

In order to determine what kinds of general educational subjects are of most worth to boys taking the all-day vocational agriculture courses in secondary schools, the questionnaire on page 13 was sent, as was described on page 12, to sixty well-known, outstanding educators in the field of general education, many of whom have taken active parts on national committees dealing with problems of secondary education and with the problems of curriculum construction in general education. The facts enumerated on the forty-seven questionnaires which were returned are shown in Table 4 following. The table shows the ten general educational subjects considered of most importance to boys enrolled in all-day vocational agriculture courses in secondary schools, arranged in order of their estimated importance:

1. English
2. Physical and health education
3. Social problems or sociology
4. Civics
5. General science
6. General mathematics
7. Biology
8. American history
9. Economics
10. World history

All these ten subjects were checked by more than fifty per cent, or in other words, by a majority, of the group which returned the questionnaires. All other subjects listed on the questionnaire were checked by less than the majority of the group.

TABLE 4

Distribution of Forty-Seven Opinions Regarding the General Educational Subjects of Most Worth to Boys Enrolled in All-Day Vocational Agriculture Courses in Secondary Schools

Subject	Persons Checking		Subject	Persons Checking	
	Number	Per Cent		Number	Per Cent
English	43	91.5	Chemistry	20	42.6
Physical and health education	42	89.4	Physics	18	38.3
Social problems or sociology	36	76.6	Physiology	9	19.2
Civics	33	72.2	Typewriting	7	14.9
General science	31	66.0	Psychology	5	10.6
General mathematics	30	63.8	Algebra	3	6.5
Biology	29	61.7	Geometry	3	6.5
American history	29	61.7	European history	3	6.5
Economics	27	57.4	Foreign languages	2	4.3
World history	25	53.2	Vocations or occupations*	2	4.3
Practical arithmetic	23	48.9	Ancient history	1	2.1

* Added to the list by two persons returning the questionnaire.

The majority opinion of the group asked to render expert opinion on the selection of these educational subjects decidedly indicates a tendency to overcome the present weaknesses of rural secondary education which have been revealed by a recent study made in this field. A quotation from this study follows:

"A critical examination of the several groups of data on the programs of studies of rural high schools indicates clearly that the core of the rural high-school curriculum is still largely composed of college preparatory subjects. It is academic in its most common features. It lacks balance, whether viewed from the angle of the common developmental needs of the adolescent, or from the angle of individual differences in interests and aptitudes, not to mention purposes, or from the angle of the social adjustments commonly needed by the great mass of people in modern life. As far as any dominant aim is apparent, it is the preparatory aim. All other aims are secondary, and they generally but surely yield to this dominant aim, as the groups considered represent schools of smaller enrollments."[1]

Some Comments Mentioned on the Questionnaires

The questionnaire returned by the forty-seven persons asked to express their opinions on the general educational subjects of most worth to farm boys enrolled in all-day vocational agriculture courses in secondary schools contained some interesting comments; also the checking done on some of these questionnaires is worth mentioning because both comments and checking bring out differing points of view. Some of these are quoted here.

1. "I have answered the questionnaire to the best of my ability, but let me make this comment. To me it seems impossible to construct an adequate program for any kind of high school, to say nothing of a high school to meet rural needs, on the basis of the conventional high school subjects as they stand today."
2. "In my judgment the required subjects, outside of agriculture and closely related science subjects, should not be more than seven units out of a total of 15 or 16. While I have checked four social science subjects, I believe these should not total more than 2 or 2½ units."
3. "I include World History if it can be taught from the standpoint of progressive development of civilization, showing how various countries and ages have contributed to our present ideas of democracy, liberty, social and religious standards, etc. Not dates, battles, rulers, etc."
4. "I include civics not as a minute study of constitutions but as the citizen's relation to law and law enforcement with respect to community, state and nation."
5. "As a rule, I am in favor of a minimum of required subjects. The larger the per cent of variables the greater the opportunity to take advantage of individual differences through guidance."
6. "Naturally a good deal depends upon the content of these courses, the objectives and aims fostered, and the instructional procedures; that is, biology is important because it gives the child orientation in the origin, nature, functions, etc., of living organisms with the view of better understanding human life. It is not a detailed study of earthworms, with emphasis on accumulating a lot of technical terms, learn-

[1] Committee on Rural Education, "The Status of Rural Education," *Thirtieth Yearbook*, Part I, National Society for the Study of Education.

ing a lot of isolated facts unrelated to life, and drilling on unimportant facts to be mastered for their own sake. Much greater emphasis should undoubtedly be given both in the vocational and the general educational subjects to such fields as sociology and economics; above all things, learning about the practical things of life."

7. "Practical arithmetic should be included as a part of general mathematics."

8. "I would not include practical arithmetic if 'shop math' is a part of the vocational work."

9. One outstanding educator checked only Economics, Biology, Physics, and Chemistry. Another bracketed Latin, Spanish, French, German, Algebra, Geometry, and General Mathematics, and wrote after them: "Absolutely *Not,* they need their *time* for things of necessity."

Still another outstanding educator wrote: "I have checked only two of the items in your questionnaire (English, and Physical and Health Education). I would certainly require some science but which one would depend somewhat upon the character of the courses in vocational agriculture and partly upon the facilities available. I would also make some requirement in the field of social studies and probably one in the field of mathematics, but just what the requirements would be would depend upon several factors."

10. Another distinguished educator checked only four subjects on the questionnaire and three of these he conditioned. He checked English and limited it to business letters; he checked Practical Arithmetic and limited it to Farm Arithmetic; and he checked Economics and limited it to Farm Economics. As the fourth subject he checked Physical and Health Education.

11. Another person checked Practical Arithmetic but would include it only in cases where it had not been taught adequately in the grades. He also checked and bracketed Biology, Physics, and Chemistry, indicating he would include them only if they were not covered in General Science.

12. One person crossed out Latin, Spanish, French, German, Algebra, Geometry, General Mathematics, Ancient History, Psychology, Physiology, and Typewriting and said: "Subjects crossed out, certainly *Not.*" The same person checked only three subjects: Practical Arithmetic, General Science, and Physical and Health Education; and further commented as follows: "All the other subjects (not crossed out) would be desirable *if* they were organized psychologically with respect to the needs of the learners. That I have not marked them to be required is due to the underlined parenthesis (as they are now commonly being taught) in your explanation on the questionnaire."

Algebra and Geometry Receive Little Recognition

It should be particularly noted that algebra and geometry were each checked by only three persons, or by 6.5 per cent of the forty-seven persons who returned the questionnaire. One must conclude from this verdict that algebra and geometry were not considered, by the persons answering the questionnaire, as very important subjects for boys enrolled in vocational agriculture courses in secondary schools and that these two very common high school subjects should not be made required subjects or "constants" in the vocational agriculture curriculum in secondary schools.

However, in making a study of the general educational subjects actually being pursued by boys enrolled in vocational agriculture courses in secondary schools, as shown in Table 27 on page 74, it is seen that, as reported by those returning the questionnaire, algebra and geometry are two of the three most common general educational subjects taken by boys enrolled in vocational agriculture courses.

Foreign Languages, Ancient History, and European History Are Ranked Low

Table 4 on page 24 shows that only two persons, or 4.3 per cent of the group of forty-seven, checked any foreign language; one person, or 2.1 per cent, checked ancient history; and three persons, or 6.5 per cent, checked European history, thus clearly indicating that these three high school subjects are considered by these expert educators of little importance to boys enrolled in vocational agriculture in secondary schools and that they should not be made required subjects in the vocational agriculture curriculum. Neither should they be suggested as elective subjects in such a curriculum.

Attitude of the North Central Association

From an article in the March, 1931, issue of the *North Central Association Quarterly*, by J. B. Edmonson, secretary of the North Central Association of Colleges and Secondary Schools, the following statements are quoted:

> 1. "The Association (North Central) has never attempted to make any prescription concerning the contents of a course that is taught. Its actions have always been purely suggestive or recommendatory. . . .

2. "It limits its activities to the academic fields. . . .

3. "I am convinced that the secondary schools have a very large amount of freedom in determining what is to be taught and how it is to be taught.

4. "The fact that certain changes have not been made in the secondary schools is traceable in my opinion to the marked conservatism of high school administrators and teachers and to the failure of advocates of new ideas to present convincing arguments. The secondary schools of the North Central territory are not in my opinion restrained by standards of our Association from the introduction of new courses or new materials."

This quotation is here introduced merely to show that the North Central Association leaves the states and the school administrators free to select the kind of vocational and non-vocational subjects they desire to introduce into the vocational course in the secondary school.

Further Discussion of This Problem

As will be brought out later in this study, the selection of general educational subjects regarded by experts in general education as most suitable for vocational agriculture pupils does not coincide with those suggested in many State Plans for Vocational Education or with the practice actually going on. Therefore this matter regarding non-vocational subjects in vocational agriculture curricula is further discussed in Part V of this study in connection with State Plans for Vocational Education; and also in Part IX, in connection with the discussion on non-vocational subjects pursued by boys enrolled in vocational agriculture courses.

Summary and Conclusions

A group of forty-seven outstanding educators in the field of general education regard the following subjects, in order given, as the ten subjects of most worth to boys enrolled in vocational agriculture courses in secondary schools:

1. English
2. Physical and health education
3. Sociology
4. Civics
5. General science
6. General mathematics
7. Biology
8. American history
9. Economics
10. World history

These same educators are almost unanimous in their opinions that such subjects as the following:

1. Ancient history
2. Foreign languages
3. European history
4. Algebra
5. Geometry

are of comparatively little worth to boys taking the vocational agriculture course in preparation for farming, and that these very common secondary school subjects should not be made required subjects or constants in vocational agriculture curricula in secondary schools.

Contrary to the belief of many, the North Central Association does not force any of the subjects last listed into the curricula of any vocational course in secondary schools. This association limits its activities entirely to the academic field and does not concern itself with any phase of vocational education. Therefore any state or any school is free to determine and to offer the subjects it desires in any of its vocational curricula.

PART V

STATE PLANS FOR VOCATIONAL EDUCATION IN AGRICULTURE

The teaching of vocational agriculture in public schools or classes in any state under the provisions of the Smith-Hughes Act is done in accordance with the standards and conditions formulated in the State Plans for Vocational Education of the State.

The formulating, in each state, of plans describing in considerable detail the kinds of vocational education in agriculture the state proposes to administer, with the use of federal funds made available in the Smith-Hughes Act, is made mandatory in the Act. This point is made clear in the following quotation taken from Section 8 of the Smith-Hughes Act:

> "That in order to secure the benefits of the appropriation for any purpose specified in this Act, the State Board (for vocational education) shall prepare plans, showing: (1) the kinds of vocational education for which it is proposed that the appropriations shall be used; (2) the kinds of schools and equipment; (3) courses of study; (4) methods of instruction; (5) qualifications of teachers; and, (6) in the case of agricultural subjects the qualifications of supervisors or directors; (7) plans for the training of teachers; and (8) in the case of agricultural subjects plans for the supervision of agricultural education as provided for in Section 10. Such plans shall be submitted by the State Board to the Federal Board for Vocational Education; if the Federal Board finds the same to be in conformity with the provisions and purposes of this Act, the same shall be approved."

The Smith-Hughes Act further states (in Section 10) the following:

> "That in order to receive the benefits of such appropriations (those available for vocational education in agriculture) for the salaries to teachers, supervisors or directors of agricultural subjects the State board of any State shall provide in its plan for agricultural education (1) that such education

shall be that which is under public supervision or control; (2) that the controlling purpose of such education shall be to fit for useful employment; (3) that such education shall be of less than college grade and (4) be designed to meet the needs of persons over fourteen years of age, (5) who have entered upon or who are preparing to enter upon the work of the farm or of the farm home; . . . (6) that such schools (schools offering vocational instruction and training in agriculture) shall provide for directed or supervised practice in agriculture, either on a farm provided for by the school or other farm, for at least six months per year; . . . ."

The Instruction and Training in Vocational Agriculture Offered in All-Day Classes in Secondary Schools and Stipulations of the State Plans

In reality the state plans for vocational education in agriculture within the state are nothing more than a set of specifications showing how the job of teaching vocational agriculture should be done. The contractual relations between the Federal Board for Vocational Education and a State Board for Vocational Education are based entirely upon the facts or specifications mentioned in the state plans. It therefore necessarily follows that the policies outlined in the state plans will be carried into operation by the various public schools within the state undertaking the work. In other words, vocational agriculture in all-day classes in secondary schools will be conducted in harmony with policies outlined in the state plans.

Because of this fact there has been included in this study an analysis of basic requirements of all-day or full-time schools or classes in vocational education in agriculture as formulated in state plans for vocational education of the various states.

A request for a copy of the State Plans for Vocational Education was sent to each of the forty-eight state supervisors of vocational agriculture. Copies of twenty-seven state plans were received. Many supervisors replied that they had no copies available for distribution. The twenty-seven copies of state plans which were received were analyzed and studied. The results are incorporated in the discussion which follows.

General Conditions Complied With

All copies of the state plans which were analyzed contained statements similar to the following as to general conditions:

1. The vocational education shall be under public supervision and control.

2. The controlling purpose of vocational education shall be to fit for useful employment.

3. The instruction and training shall be of less than college grade.

4. The instruction and training shall be for persons over fourteen years of age.

5. Every dollar of federal funds expended shall be matched by a dollar of state or local money or of state and local money combined.

State Plans Make Provision for Instruction in Vocational Agriculture in All-Day Classes in Public High Schools

Each of the twenty-seven copies of state plans studied made provision for offering vocational agriculture in all-day classes in secondary schools. With few exceptions the work in vocational agriculture is organized as a department in regular high schools, including county high schools, union high schools, consolidated high schools, and rural consolidated vocational schools.

The pupils enrolled in vocational agriculture courses in these schools take the instruction in vocational agriculture; in closely related subjects, such as farm mechanics, farm shop work; and sometimes, in related science, under a specially trained teacher of vocational agriculture, who also is in charge of, and oversees, the supervised farm practice work of the pupils. The non-vocational subjects are taken in regular high school classes with other pupils enrolled in the school.

School Time to Be Given to Vocational Agriculture as Specified in State Plans for Vocational Education

The amount of school time as specified in state plans to be given to vocational agriculture in all-day classes in secondary schools varies. Table 5 shows that twenty out of the twenty-seven state plans studied specify that the vocational agriculture courses in high schools may be from two to four years in length.

The amount of school time to be given daily to vocational agriculture subjects varies, as is shown in Table 6, from ninety minutes to one-half a school day. Ninety minutes per day is the prevailing time allotment.

TABLE 5

LENGTH OF THE ALL-DAY VOCATIONAL AGRICULTURE COURSE IN SECONDARY SCHOOLS

As Specified in Twenty-Seven State Plans for Vocational Education

Length of the Vocational Agriculture Course	Number of States Specifying Length
Two to four years	20
One to four years, depending on time necessary to train a boy for the business of farming	1
Four years	2
Such as will meet the needs of pupils in the community	1
Four years except where otherwise specified	1
Two to three years	1
Usually two years in length	1
Total	27

In addition to this school time to be devoted to vocational work, all state plans specify that each pupil enrolled in vocational agriculture must also engage in directed or supervised farm practice work for a period of not less than six months per year.

TABLE 6

DISTRIBUTION OF REGULAR SCHOOL TIME TO BE GIVEN, PER DAY, TO VOCATIONAL AGRICULTURE AND TO CLOSELY RELATED WORK

As Specified in Twenty-Seven State Plans for Vocational Education

Amount of School Time	Number of State Plans Specifying Amount
At least one double school period	1
Ninety minutes daily	10
Not less than 90 minutes per day	3
Ninety to 135 minutes per day	2
Ninety to 180 minutes per day	4
One-half a school day during first and second school years and 90 minutes daily during the third and fourth school years	2
One-half a school day daily to vocational subjects and to related science subjects taught by the vocational agriculture instructor	2
One-half a school day daily to strictly vocational work	3
Total	27

Courses of Study in Vocational Agriculture as Outlined in State Plans

Statements in state plans regarding courses of study in vocational agriculture for all-day pupils in secondary schools, and type or illustrative curricula vary greatly. Pages 35 to 39 show four such curricula; the first two differ decidedly from the last two, in that they are more vocational and less academic, in that they put a greater emphasis upon the vocational work, and in that they suggest a more practical type of general educational work.

Most of the state plans examined, however, do not outline curricula; they merely describe the course of study to be offered in vocational agriculture. The following quotations taken from various state plans are typical descriptions:

"Each school shall emphasize instruction and farm practice work in animal and crop enterprises most common and most profitable on the farms of the pupils enrolled. The work must be especially designed to train practical farmers."

"The controlling purpose of the course of study in vocational agriculture is to prepare the student for efficient farming, intelligent participation in the affairs of the agricultural group, and his functioning as a desirable American citizen. The course of study will also give excellent preparation for the college of agriculture and for colleges not requiring foreign languages for entrance."

"There is no state-wide curriculum for students taking vocational education in agriculture in secondary schools. Academic and related courses which seem most nearly to fit the needs of the class or the individual student are chosen from the program of studies offered by the secondary school. The local supervisor of vocational agriculture, and the local teacher of vocational education in agriculture, coöperate in making up a curriculum of each class or student. In each case it is intended to make up a well-balanced curriculum which will prepare for good citizenship and for other social obligations and responsibilities of farmers. There is no state-wide course of study in agriculture. Enterprises are selected to make up courses of study and emphasis is determined on the basis of a careful survey of the farming of the community, and the needs of the students taking the courses. Enterprises and jobs which receive great emphasis in one region may be entirely omitted in another."

"The course of study in agriculture is made up of the different farm enterprises carried on in farming in the community with some emphasis on potential enterprises that may have future importance. Each teacher of agriculture with the approval of his local superintendent and the state office prepares his course of study for high school pupils. The supervised practice work of the pupils is made up of the same enterprises that are given in the course of study."

"The four year curriculum shall usually consist of one year each of Plant Husbandry, Animal Husbandry, Farm Mechanics, and Farm Economics. The subject matter in each year's work shall be determined by the needs of the farm boys of the community."

State Plans and General Educational Subjects

The state plans examined in this study give too little attention, even in way of suggestion, to the kind of general educational subjects which would be of most worth to boys enrolled in vocational agriculture courses in secondary schools. The two curricula beginning near the bottom of page 37 and continuing to page 39 are typical of the majority of curricula pursued by boys enrolled in vocational agriculture courses. Both are nothing more than traditional academic, college preparatory curricula with vocational agriculture included as an elective subject. The general educational subjects listed in them do not coincide very closely with the opinions of educators in the field of general education as given in Part IV of this study. Certainly they are not the best curricula for boys taking the vocational agriculture courses in secondary schools in order to fit themselves for useful farm employment and for the other responsibilities and social and civic activities of the farmer's life.

TYPE FOUR-YEAR CURRICULUM IN VOCATIONAL AGRICULTURE FOR HIGH SCHOOLS[1]

At least fifty per cent of the school time of each student in all-day courses is employed in the study of agriculture. Supervised home project work or other approved farm work is carried on in all cases in connection with the classroom instruction.

Academic and Related	*Vocational*
	FIRST YEAR
400 to 750 minutes per week	600 to 750 minutes per week
English *General mathematics *General science	Operative and managerial training in selected plant production enterprises adapted to the community such as corn, potatoes, pastures, apples, etc., including related soil and fertilizer management, the control of insects, diseases, and weeds injurious to the crops, etc. Shop work and farm machinery.

[1] Taken from a State Plan for Vocational Education.

	SECOND YEAR
400 to 750 minutes per week English *Biology *Modern history (1700 to present)	600 to 750 minutes per week Operative and managerial training in selected animal production enterprises adapted to the community, such as dairying, poultry, swine, horses, etc., including related work dealing with feeds, breeding, diseases, parasites, etc., and shop work involving wood, iron, concrete, rope, and leather.
	THIRD YEAR
400 to 750 minutes per week English *American history and civics *Physics	600 to 750 minutes per week Farm Management Case studies of farms exemplifying the principal types of farming of the community such as dairy farm management, dairy-fruit farm management, poultry farm management, etc. Continued study of the major production enterprises in relation to managerial problems encountered in the case studies. Farm shop related to the major enterprises and to the farm as a whole.
	FOURTH YEAR
400 to 750 minutes per week English	600 to 750 minutes per week Farm Management Continued study of case farms exemplifying various types of farm organization of the community with emphasis upon farm law, records and accounts, rural, economic, and social problems, etc. Farm shop emphasizing water supply systems, lighting systems, plumbing, farm buildings, not before considered, etc.

VOCATIONAL EDUCATION IN A HIGH SCHOOL AGRICULTURE DEPARTMENT[2]

FOR DAY PUPILS SHOULD HAVE TWO PARTS

Part 1. Vocational Education *50 per cent of pupil's time*	*Part 2. General Education* *50 per cent of pupil's time*
1. General survey of agriculture (10 per cent) 2. Project study and project work (40 per cent) centering on:	Cultural and good citizen training, selected from one or more of the regular high school courses, and dealing with such subjects as:—

* Three subjects in addition to agriculture may be carried by students who rank or who give evidence of ability to rank in the upper fourth of their classes in secondary or intermediate school. Two subjects are carried by students who do not meet this standard.

[2] Taken from a State Plan for Vocational Education.

A. Projects of the pupils *a*. At home, as a rule *b*. Near home, occasionally *c*. Pupil responsible, but supervised by instructor	English, every year Social science, including community civics and economics
B. Projects of the department *a*. At high school, rarely *b*. Neighborhood demonstrations, as of pruning, spraying, hotbed making, or greenhouse work	Natural science, including elementary science, biology, physics, and chemistry Drawing, freehand and mechanical
C. Substitutes for projects *a*. Work on approved farms with agreed-upon educational duties, as cost-accounting one or more cows or one or more crops *b*. Employer chiefly responsible but supervision by instructor	Shop work Business, including typewriting, business forms and filing, bookkeeping, commercial geography, and commercial law
D. Remark.—The agriculture instructor must, as a rule, assume full responsibility for teaching the "related study" required for the proper understanding and execution of the projects of his pupils. He must generally teach his boys the vital correlation between their projects and such subjects and activities as arithmetic, biology, physics, chemistry, entomology, drawing, shop work, accounting, filing, farm journal reading, agricultural economics, accident prevention, and "first aid."	Physical training, personal hygiene, accident prevention, and "first aid." Music Recreation

VOCATIONAL AGRICULTURE COURSE[3]

(R-Required E-Electives)

		Value
First Year		
English	R	1
General science	R	1
Agriculture (plant husbandry)	R	1½
History	E	1
Mathematics	E	1
Manual arts	E	½
Second Year		
English 2	R	1
Agriculture (animal husbandry)	R	1½

[3] Taken from a State Plan for Vocational Education.

Biology	R	1
History 2	E	1
Mathematics 2	E	1
Third Year		
English 3	R	1
U. S. history	R	1
Mathematics 3	E	1
Language	E	1
Science 3-4	E	1
Manual arts	E	½
Agriculture (community specialties)	E	1½
Fourth Year		
English 4	R	1
Civics and economics	R	1
Language	E	1
Science 4-3	E	1
Mathematics	E	1
Agriculture (community specialties)	E	1½

In addition to the 90 minutes per day required for agriculture in the classroom, provision shall be made for equivalent of 90 minutes per day for a period of at least six (6) months of supervised farm practice, such as home projects and other farming activities.

TYPE COURSE OF STUDY FOR VOCATIONAL AGRICULTURE[4]

Outline of Course	*Length of Subject in Weeks*	*Periods per Week*	*Minutes per Period*
First Year			
1. Farm shop	36	5	90
2. General science	36	7	45
3. Mathematics — general mathematics or algebra	36	5	45
4. English	36	5	45
Second Year			
1. Crop production	36	5	90
2. History (modern)	36	5	45
3. Mathematics (plane geometry)	36	5	45
4. English	36	5	45
Third Year			
1. Animal production	36	5	90
2. American history and civics	36	5	45
3. Elective—solid geometry, biology, or physics			
4. English	36	5	45
Fourth Year			
1. Farm management and marketing	36	5	90
2. Economics and social sciences	36	5	45

[4] Taken from a State Plan for Vocational Education.

3. Elective—physics, chemistry, etc.	36	7	45
4. English	36	5	45

Note: The above course of study should be considered as only tentative. Changes and improvements may be made from time to time when such changes can be made advantageously and are in keeping with new and desirable information relative to curriculum building.

Each school shall emphasize instruction and farm practice work in animal and crop enterprises most common and most profitable on the farms of the pupils enrolled. The courses shall include provisions for instruction in English, citizenship, social sciences, mathematics, and such coördinating subjects as general and technical sciences.

Methods of Instruction as Described in State Plans

Statements in state plans describing methods of instruction which are to be followed in teaching all-day pupils enrolled in vocational agriculture classes in secondary schools vary greatly. The following are illustrations taken from state plans:

1. "The methods of instruction shall combine supervised study, laboratory demonstrations, recitations and field trips, supplemented by individual practicums and home projects. The development shall be from the standpoint of productive farming and not from that of scientific investigations."
2. "The methods of instruction will vary depending upon the kind of farm job being taught. Use will be made of the recitation, supervised study, demonstration, field trips, laboratory and individual supervised practice work."
3. "The methods of instruction shall be such as to prepare the pupils for the business of farming."
4. "The primary aim of vocational education in agriculture is to prepare students for or to assist them in productive and profitable farming. Instruction in the practical problems encountered in the daily work of the farm is covered by such methods and devices as technical study, laboratory tests, surveys, and by practical work on the students' home projects. Students are brought into contact with the best agricultural practice of the community and with the men who are carrying on the most successful farm business. The home project and other forms of supervised practice work are used to give students operative and managerial experience. . . . With many types of work group instruction is effective and economical. This is particularly true of work involving managerial considerations. Groups will frequently do better work than individuals where the object of the work is to have students obtain training in the processes of considering and evaluating factors in making decisions. . . ."
5. "Lack of farm land and of elaborate special agricultural

laboratory and shop equipment at the high school makes group teaching difficult in both the laboratory and the practical aspects of the training of agriculture department pupils. The problem of the department (of vocational agriculture) is to develop its headquarters into a combined study room and agricultural science laboratory, and, at school and on neighboring farms, to provide for a fair proportion of group instruction. The department may be expected to excel principally in the educational benefits which follow individual methods. The department has no extensive farming operations on its premises. . . . Its fundamental requirement is that all applicants prior to admission must make arrangements for projects, or acceptable substitutes, at or near home. Thus the interest of both instructor and pupils tends to be not centripetal, but centrifugal—not inward toward things done at school so much as outward toward the separate projects and substitutes for project responsibilities which are being carried by the individual pupils at or near their several homes. For this type of department no methods not in keeping with those just mentioned will be approved by the State Department as either efficient or economical."

6. "The methods of instruction should be such as best prepare the pupils for the occupation of farming. Instruction shall be given on the farm job basis and shall consist of discussions, laboratory and practical field work, project work and other forms of supervised practice work. The order of presentation should depend upon the seasonal importance of the work and the needs of the boys in the class and not upon the order in a text. The aim in organizing the subject matter of the vocational agriculture courses should be to bring together in a usable form all that related information which the trainee must have and use in order to be successful with the important crops and types of livestock in the community and which helps most in his supervised practice work."

Summary and Conclusions

The sections of twenty-seven State Plans for Vocational Education pertaining to instruction in all-day schools and classes in vocational agriculture in secondary schools were analyzed as outlined in the preceding discussion.

With very few exceptions, it appeared to the writer that all the statements and suggestions made in most of these state plans with reference to

1. the purpose of the vocational education in agriculture in all-day schools or classes,

2. the courses of study, and
3. the methods of instruction

were entirely too generalized and too vague; moreover, they were often inefficient and entirely lacking in direct and specific statements as to what should characterize this type of work in the secondary schools in order to obtain efficient results. Little or no reference was made to such important points as are implied by the factors mentioned in Part III of this study, many of which, in the opinion of the writer, need to be forcibly stressed if an effective training program is to be followed.

Furthermore, they give practically no suggestions and offer no guidance in the selection of those general educational subjects which should form a part of the vocational agriculture curriculum in order to promote the most efficient type of rural citizenship.[5]

[5] It should be mentioned in this connection that a few states supplement their state plans for vocational education in agriculture with special circulars in which the meager and general statements regarding the provisions of the state plans are enlarged upon, made more specific, and are amply illustrated. In other ways, also, such circulars, published under some such title as "Vocational Agriculture in High Schools," make clear to local school authorities all the underlying purposes of vocational education in agriculture, and suggest effective ways and means of attaining these purposes. Such a procedure seems highly essential for the attainment of the most efficient results in any state.

PART VI

TYPE OF BOYS ENROLLED IN ALL-DAY VOCATIONAL AGRICULTURE CLASSES IN SECONDARY SCHOOLS

Information concerning the type of boys enrolled in all-day vocational agriculture classes in secondary schools was secured from eighty-eight teachers of vocational agriculture, located in twenty-eight different states, and from not more than five teachers in any one state. Of these twenty-eight states five are located in the eastern part of the United States, seven in the mid-west, nine in the south, and seven in the west.

This first part of this study, as shown in Table 7, considers 2,740 boys enrolled in all-day vocational agriculture classes in secondary schools during the school year 1930-1931.

The information secured concerning these boys was as follows:

1. The purpose for which the boys are taking the work in vocational agriculture.
2. The ages of the boys enrolled in all-day vocational agriculture classes in secondary schools.
3. The residences of these boys; that is, numbers living on farms, and not on farms.
4. The place where these boys do their supervised farm practice work.

Purpose for Which Boys Enrolled in All-Day Vocational Agriculture Classes in Secondary Schools Are Taking the Vocational Work

Table 7 shows that 68.7 per cent of the boys enrolled in all-day vocational agriculture classes, as reported by eighty-eight teachers of vocational agriculture, are taking the work in vocational agriculture for preparation for the pursuit of farming as a means of livelihood; 13.7 per cent are taking the work for some other purpose; and 17.6 per cent are in doubt as to the purpose for which they are taking the work.

The data in columns three and four in Table 7 would seem to

TABLE 7

DISTRIBUTION OF NUMBERS OF BOYS ENROLLED IN ALL-DAY VOCATIONAL AGRICULTURE CLASSES ACCORDING TO THE PURPOSE FOR WHICH THEY ARE TAKING VOCATIONAL AGRICULTURE

As Reported by Eighty-Eight Teachers of Vocational Agriculture

School Year	Number Taking the Work for Preparation for Farming	Number Taking the Work for Purposes Other Than Preparation for Farming	Number in Doubt as to Purpose	Total
First	862	162	247	1,271
Second	612	110	127	849
Third	332	81	79	492
Fourth	77	23	28	128
Total	1,883	376	481	2,740
Per cent	68.7	13.7	17.6	100.0

reveal a weakness of the vocational training program in agriculture in our secondary schools. "The aim of the all-day vocational agriculture course offered in secondary schools is to train prospective farmers for proficiency in the farming occupation in which they are fitting themselves to engage." Admitting to such classes boys who are working for some other purpose, or who are in doubt as to their purpose is therefore a violation of this aim. The weakness here referred to can be eliminated by teachers of vocational agriculture and by school administrators by making clear to a school community and strictly adhering to the principles that "the instruction and training in vocational agriculture is given to a selected group of pupils who need it, want it, and are able to profit by it." Ninety and two-tenths per cent of the educational experts in vocational education, returning the questionnaire on what constitutes an effective vocational training program in vocational agriculture, agreed that the aim stated in the previous paragraph should dominate the carrying out of the purpose of vocational education in agriculture in all-day schools or classes. This particular point is here reiterated merely to emphasize that an effective vocational education course has a decidedly specific aim or function, and that pupils not interested in the work for which such a course is giving training should not be allowed to enroll in it. Allowing such enrollment would involve a social waste.

The Age of Boys Enrolled in All-Day Vocational Agriculture Classes in Secondary Schools

Facts gathered from eighty-eight teachers of vocational agriculture, as indicated in Table 8, show that the median ages of 2,809 boys enrolled in all-day vocational agriculture classes in secondary schools are as follows:

In the first year class .... 14.9 years
In the second year class .... 15.9 years
In the third year class .... 16.8 years
In the fourth year class .... 17.5 years

The table further shows that the median age of the entire group of 2,809 boys is 15.7 years. From the data in Table 8 was also calculated the median age of the 2,204 boys in the first and second year classes, or 78.5 per cent of the entire group enrolled in the first and second year classes. The median age of this group was found to be 15.3 years.

TABLE 8

DISTRIBUTION, ACCORDING TO AGE, OF BOYS ENROLLED IN ALL-DAY VOCATIONAL AGRICULTURE CLASSES IN SECONDARY SCHOOLS

As Reported by Eighty-Eight Teachers of Vocational Agriculture

Age	Number of Boys in First Year Classes	Number of Boys in Second Year Classes	Number of Boys in Third Year Classes	Number of Boys in Fourth Year Classes	Total	Per Cent
13 to 14	274	26	0	0	300	10.7
14 to 15	447	154	6	0	607	21.6
15 to 16	320	268	85	11	684	24.4
16 to 17	177	211	167	32	587	20.9
17 to 18	71	155	114	43	383	13.6
18 to 19	28	44	70	30	172	6.1
19 and up	8	21	33	14	76	2.7
Total	1,325	879	475	130	2,809	
Per cent	47.2	31.3	16.9	4.6		100.0
Median age ...	14.9	15.9	16.8	17.5	15.7	

Table 8 further shows the percentage distribution of these boys, according to age, to be as follows:

13 to 14 years of age	10.7 per cent
14 to 15 years of age	21.6 per cent
15 to 16 years of age	24.4 per cent
16 to 17 years of age	20.9 per cent
17 to 18 years of age	13.6 per cent
18 to 19 years of age	6.1 per cent
19 and over years of age	2.7 per cent
Total	100.0 per cent

According to the data in Table 8, approximately three-fourths (78.5 per cent) of the 2,809 boys enrolled in the all-day vocational agriculture classes studied were in the first two year classes and 21.5 per cent of the boys were in the third year and fourth year classes.

According to the opinion of 92.1 per cent of the persons returning the questionnaire concerning factors which characterize an effective training program in vocational agriculture, as shown in Table 2 on pages 16 to 17, the following factor is an outstanding characteristic of such a program: "The instruction and training in vocational agriculture is given to a selected group of pupils who need it, want it, and are able to profit by it."

In the light of this outstanding characteristic the question at once arises: Are the 56.7 per cent of boys between 13 and 16 years of age old enough to profit much by real vocational education? The same question also presents itself regarding the majority of the 78.5 per cent of the boys enrolled in the first two year classes in vocational agriculture, where the median age of the boys enrolled in the first two year classes is 15.3 years. Pupils under 14 years of age should never be enrolled in federally-aided vocational classes, because the Smith-Hughes Act itself, as is mentioned on page 31, fixes the minimum age of entrance at 14 years.

Where Boys Enrolled in All-Day Vocational Agriculture Classes in Secondary Schools Live

In practically every instance high schools offering instruction in vocational agriculture have little or no farm land, livestock, and farming equipment of their own; therefore they must make provision for supervised farm practice (training experience) in the form of home projects conducted on the home farms of the pupils enrolled in the vocational agriculture classes. In the execution of these projects, which invariably take the nature of a farm enterprise such as pork production, egg production, potato production, corn production, and in the performance of other supplementary farm practice work, boys enrolled in vocational agriculture classes are provided with real farming experiences, both operative and managerial, through which they acquire operative skill and managerial abilities in farming in a real occupational environment. These points are mentioned here merely to show that where boys enrolled in vocational agriculture classes live is an important factor in vocational education in agriculture in secondary schools.

Table 9 shows that a large majority of the boys, 81.3 per cent, of the group reported upon live on farms, a fact which indicates that they have opportunities for engaging in effective supervised farm practice work. Table 10 shows where boys enrolled in all-day vocational agriculture classes in secondary schools actually perform their project work. This latter point is discussed in detail on pages 47 and 48.

Because of the very extensive use of home projects as a means of providing participating (training) experiences in farming, it is very essential that pupils enrolled in vocational agriculture classes in secondary schools live on a farm, or that they have other adequate farm facilities for engaging in supervised farm practice work, in order to make sure that "the vocational training environment is the working environment itself or a replica of the working environment."

TABLE 9

DISTRIBUTION OF NUMBERS OF BOYS IN ALL-DAY VOCATIONAL AGRICULTURE CLASSES LIVING ON FARMS AND NUMBERS NOT LIVING ON FARMS

As Reported by Eighty-Eight Teachers of Vocational Agriculture

School Year	Number of Boys Living on Farms	Number of Boys Not Living on Farms	Total
First	1,037	261	1,298
Second	707	167	874
Third	417	78	495
Fourth	109	17	126
Total	2,270	523	2,793
Per cent	81.3	18.7	100.0

The fact that supervised farm practice work is a necessary part of the training program in vocational agriculture is recognized in the Smith-Hughes Act, which, as was mentioned on page 31, specifies that schools giving instruction in vocational agriculture under the provisions of the Act, "shall provide for directed or supervised practice in agriculture, *either on a farm provided for by the school* or *other farm*, for at least six months per year." Similarly, all State Plans for Vocational Education make the same requirement, as was mentioned on page 33.

Where Boys Enrolled in All-Day Vocational Classes in Agriculture Conduct Their Supervised Farm Project Work

Table 10 shows where 2,772 boys enrolled in all-day vocational agriculture classes conduct their supervised farm practice work. The fact that 78.5 per cent of these boys conduct this work on the home farm or on some other farm speaks well for this type of work.

TABLE 10

DISTRIBUTION OF NUMBERS OF BOYS ENROLLED IN ALL-DAY VOCATIONAL AGRICULTURE CLASSES ACCORDING TO WHERE THEY CONDUCT THEIR SUPERVISED FARM PROJECT WORK

As Reported by Eighty-Eight Teachers of Vocational Agriculture

School Year	On Home Farm	On Farm Other Than Home Farm	On Vacant Lot	On Back-yard in Town	Number Having No Project	Number Substituting Farm Experience	Total
First	1,030	70	37	109	51	13	1,310
Second	661	45	28	81	30	8	853
Third	397	22	22	32	9	2	484
Fourth	89	9	1	6	17	3	125
Total	2,177	146	88	228	107	26	2,772
Per cent	78.5	5.3	3.2	8.2	3.9	.9	100.0

That "the vocational training environment is the working environment itself, or a replica of the working environment" is a factor characterizing an effective training program in vocational agriculture. This factor was designated as essential by 90.2 per cent of the persons answering the questionnaire concerning factors characterizing effective vocational education, as shown on page 16. In farming the working environment is a farm; therefore if agricultural training is to be effective, that training (practical experience in farming) should be given on a farm.

Vacant lots and back yards are inappropriate places to give effective training for most types of farming. They are not farms. They are absolutely lacking in farm environment. They are in no adequate or thorough way a substitute for farms as a place affording opportunity for real, practical farm training.

Another point, also, should be emphatically mentioned here. Boys having no projects and doing little or no supervised farm practice work are getting very poor and inadequate preparation for farming; they are not being vocationally trained.

Table 23 on page 68 shows that 76.1 per cent of the teachers of vocational agriculture questioned as to the dominant point of view they held regarding the purpose and value of the supervised home project work conducted by boys enrolled in all-day vocational agriculture classes, regard such work "as the important training device by means of which the pupils will acquire experience in the management of farm enterprises and skill in the execution of the various jobs involved in such enterprises." This point of view necessitates real projects, conducted on farms —that is, in a real working environment.

Summary and Conclusions

The median age of 2,809 boys, considered in this study, enrolled in all-day vocational agriculture classes in secondary schools is 15.7 years. The data on page 44 show that 78.5 per cent of these boys were enrolled in the first two year classes, 16.9 per cent in the third year class, and only 4.6 per cent in the fourth year class. The data further show that 56.7 per cent of the total number of boys are between the ages of 13 and 16 years.

These findings raise some important questions, namely: (1) Would vocational education in agriculture render a greater social service if the 78.5 per cent of the boys were in the last two year classes instead of the first two year classes? (2) If only 4.6 per cent of the boys are in the fourth year class, what happens to the 95.4 per cent of the boys found in the first three year classes? (3) Are the 56.7 per cent of the boys between the ages of 13 and 16 years old enough to profit much by real vocational education? Because of a lack of reliable information, these questions cannot be positively answered. They suggest problems for future study.

The fact that 81.3 per cent of 2,793 boys considered live on farms clearly indicates that vocational education in agriculture in secondary schools is to-day effectively reaching farm boys.

The fact that 83.8 per cent of 2,772 boys, considered in this study, conduct their supervised farm project work on farms seems to indicate that a very large percentage of boys enrolled in all-day vocational agriculture classes are receiving training in a real working environment.

Since all vocational training courses are for a selected group who need, want, and can profit by the instruction, boys not interested in preparing themselves for farming should not be enrolled in vocational agriculture classes. It is a waste of time, energy, and money to give any individual vocational training which he does not need or want, and from which he cannot profit.

Preparing boys for farming means that real farming must be a part of their training, and real farming is impossible without a farm. Unless boys expecting to enroll in vocational agriculture live on farms or have other real facilities for supervised farming, they should not be allowed to enroll in vocational agriculture classes.

A highly selected group of the right kind of boys in all-day vocational agriculture classes in secondary schools is impossible unless those responsible for the program clearly understand the aims and functions of vocational education in agriculture and make a conscientious effort to carry out these aims and functions.

The writer believes that the existing conditions concerning the points discussed in this part of this study need material improvement. That improvement can be attained by the states' giving more attention to these points in their State Plans for Vocational Education, presenting them more definitely and more specifically. Illustration 5 on page 39 is an excellent statement, taken from a state's plan for vocational education, illustrating this particular point.

PART VII

THE ALL-DAY VOCATIONAL AGRICULTURE COURSE IN SECONDARY SCHOOLS

This part of this study is based on the facts assembled from eighty-eight teachers of vocational agriculture in twenty-eight different states in different sections of the United States. The facts were secured on the same questionnaire as were those discussed in Part VI.

The particular points made the basis of investigation of this section are:

1. What is the length of the all-day vocational agriculture course in secondary schools?
2. Are the vocational agriculture subjects offered each year or in alternate years?
3. How much regular school time, per week, is devoted to vocational agriculture?
4. What is the length of the school period set aside per day for vocational work in agriculture?
5. What conditions govern admission into vocational agriculture classes?
6. What agricultural subjects are pursued by boys enrolled in the vocational agriculture classes?

The Length of the All-Day Vocational Agriculture Course in Secondary Schools

All-day vocational agriculture courses in secondary schools vary in length as shown by Table 11. The common lengths are two, three, and four school years of approximately nine months each. The length of the course is governed partly by the specifications in State Plans for Vocational Education. This point is illustrated in Table 5 on page 33, and shows that twenty out of the twenty-seven state plans studied specify that the all-day vocational agriculture course in secondary schools shall be two to four years in length. The other factors controlling the length of the course are the number of farm boys attending high school and the interest taken in the work.

TABLE 11

LENGTH, IN YEARS, OF THE ALL-DAY VOCATIONAL AGRICULTURE COURSE IN SECONDARY SCHOOLS

Length of Vocational Agriculture Course	Schools Reporting	
	Number	Per Cent
One year	0	0
Two years	27	30.7
Three years	28	31.8
Four years	33	37.5
Total	88	100.0

Ideally, the length of any preparatory vocational agriculture training course should be the time necessary to prepare the learners efficiently for the occupation in which they are fitting themselves to engage; or, as mentioned by Factor 12 on page 11, "The vocational instruction and training is continued until the trainee possesses a degree of vocational competency sufficient to enable him to secure and hold employment in the occupation in which he is fitting himself to engage." This factor was checked as a desired essential by thirty-nine of the fifty-one persons who returned the questionnaire concerning factors characterizing an effective training program in all-day vocational agriculture classes in secondary schools as shown in Table 2, which appears on page 17.

This ideal condition, however, is very difficult to secure at present. One reason for this difficulty is the fact that the all-day vocational agriculture course, in practically all instances, is tied up closely with, and is a part of, a prescribed four-year high school course. While this condition prevails, secondary schools will continue to give a two-, a three-, or a four-year course in vocational agriculture, the actual length depending upon the interest taken in the work. Factor 13, "The administration of the all-day vocational agriculture course is elastic and not rigid or highly standardized," which is discussed at length in Part X of this study, further discusses this particular problem from a slightly different point of view.

But the more fundamental reason is that nowhere do we yet possess adequate "job analyses of the many vocations of farming or of the competencies needed in them; as for example, by a man of 26 just renting or inheriting a farm," in one of the many types of farming.

It should be mentioned, however, that one state in its State Plans for Vocational Education, as shown in Table 5 on page 33, specifies that the length of the all-day vocational agriculture course shall vary, depending upon the time necessary to train an individual boy for the business of farming. Another state plan specifies that the length of the course shall be such as will meet the needs of the pupils of the community.

Vocational Subjects That Are Offered Each Year, and in Alternate Years

Some secondary schools offer the vocational agriculture subjects each year; and others offer them on an alternating basis. Table 12 shows that the latter scheme predominates. The plan of combining pupils into groups and alternating the agriculture

TABLE 12

NUMBER OF SECONDARY SCHOOLS OFFERING THE VOCATIONAL AGRICULTURE SUBJECTS EACH YEAR AND THE NUMBER OF SCHOOLS ALTERNATING THESE SUBJECTS

How Offered	Schools	
	Number	Per Cent
All vocational subjects offered each year	24	27.3
All vocational subjects offered alternate years	42	47.7
Third and fourth year subjects alternated	6	6.8
Second and third year subjects alternated	16	18.2
Total	88	100.0

subjects makes for larger classes in small schools, and at the same time enables the school with only one teacher of vocational agriculture to give a longer agricultural course than it could if one teacher had to teach each agricultural subject in the course each year. This question was studied merely to discover the practice in this respect.

The Amount of Regular School Time Devoted Weekly to Vocational Work

The amount of regular school time set aside per week for all aspects of vocational work, such as class discussion, laboratory and field work, project study and planning, farm shop work and supervised study, varies greatly, as shown by the data in Table 13. The majority of the schools studied set aside 7.5 hours per week for the work, which is the equivalent of 90 minutes per day. This allotment does not include the time devoted to supervised farm practice work or project work, which is done outside of school, and is continued by each pupil enrolled in vocational agriculture classes for a period of not less than six months per year, as specified in all State Plans for Vocational Education. This last point was discussed on page 33. Since this particular problem is intimately connected with the next problem, "The length of the school period set aside per day for vocational work in agriculture," further discussion of the question of regular school time devoted to vocational work will be taken up with the discussion of the latter problem.

TABLE 13

DISTRIBUTION OF HOURS PER WEEK OF REGULAR SCHOOL TIME DEVOTED TO VOCATIONAL AGRICULTURE IN SECONDARY SCHOOLS

As Reported by Eighty-Eight Teachers of Vocational Agriculture

School Year	Hours per Week Devoted to Vocational Agriculture *Classified by Schools*				Total
	7.5 *	7.6–10.0	10.1–12.5	12.6–15.0	
First	44	18	7	16	85
Second	43	17	7	10	77
Third	34	11	2	6	53
Fourth	20	1	1	3	25
Total	141	47	17	35	240
Per cent	58.8	19.6	7.1	14.6	100.0

* The specified minimum in state plans for vocational education in agriculture.

The Length of the School Period Set Aside, per Day, for Vocational Work in Agriculture

The amount of school time set aside, per day, for the various aspects of vocational work varies as indicated in Table 14.

Ninety minutes per day is the prevailing amount. This factor is vitally influenced by the requirements specified in State Plans for Vocational Education, as is shown in Table 6, on page 33.

TABLE 14

DISTRIBUTION OF LENGTHS OF THE SCHOOL PERIOD DEVOTED TO STRICTLY VOCATIONAL AGRICULTURE WORK IN SECONDARY SCHOOLS

Schools	Length of the Vocational Agriculture Period in Minutes				Total
	90 *	91–120	121–150	151–180	
Number of schools reporting.........	60	11	7	8	86
Per cent	69.8	12.8	8.1	9.3	100.0

* The specified minimum in state plans for vocational education in agriculture.

Since the requirements for any particular kind of vocational work specified in State Plans for Vocational Education must be met by schools coöperating with a State Board for Vocational Education, it follows naturally that local school boards will comply with these requirements. If state plans specify that 90 minutes per day shall be devoted to vocational agriculture, then that amount of time will be so devoted. If state plans specify that not less than 90 minutes per day shall be devoted to the vocational work, then most schools will devote 90 minutes to the work. The point illustrated here is that the State Boards for Vocational Education in their state plans, in practically all instances studied, fix the amount of school time devoted per day to vocational work.

How much school time per day should be devoted to vocational work is always a question of much heated discussion among educators. According to Table 6 on page 33, three of the twenty-seven state plans studied specify that one-half a school day shall be devoted daily to strictly vocational work; two specify that one-half a day shall be devoted daily to vocational work and related sciences; and two specify that one-half a day shall be devoted daily to such work during the first two years of the course and less during the third and fourth years.

In consideration of the various aspects of the vocational agriculture training program such as class discussion, laboratory and field work, analysis of farm problems, demonstrations, project study and planning, farm shop work, and supervised study, it appears to the writer that 90 minutes per day is entirely too little time for successful training of this kind. He further believes that 135 minutes or 180 minutes per day could very profitably be used.

In regard to this particular point, one outstanding authority and writer in vocational education has this to say:

> "To the writer it seems strongly probable, as stated elsewhere, that as experiences accumulate and are subjected to accurate testing it will be found that effective vocational education is to be accomplished only under conditions of concentration analogous to those required of workers engaging definitely in full time productive work. If a boy has 1500 or 1800 hours available for purposes of vocational education in a farming calling or one strongly marked strand of a farming calling, it is probable that many times as much power will be gained by concentrating them within one year of strictly vocational work at the rate of five or six hours daily for 300 days (or 365 days in the case of livestock projects) rather than at the rate of two hours a day for 200 days during four years while giving needed attention to parallel studies so engrossing as foreign language, mathematics and history."[1]

Conditions Governing Admission into Vocational Agriculture Classes

According to the data in Table 15, assembled from the responses to the questionnaire made by eighty-five teachers of vocational agriculture, in 40 per cent of secondary schools having departments of vocational agriculture any pupil enrolled in the school may elect vocational agriculture if he so desires.

On the other hand, in 40 per cent of the eighty-five departments of vocational agriculture reported upon, enrollment in vocational agriculture classes is limited to those pupils who are preparing themselves for farming and who have facilities to do adequate supervised practice work including projects.

Because of the purposes for which the work in vocational agriculture was inaugurated, as discussed in Part I of this study, the writer believes that the condition referred to in the previous paragraph should govern enrollment in vocational agriculture classes in secondary schools. Factor 2, mentioned in Table 2 on page 16, reads as follows: "The instruction and training in

[1] Snedden, David, *Vocational Education*, p. 187.

TABLE 15

DISTRIBUTION OF FREQUENCIES SHOWING THE DOMINANT SITUATION REGARDING ENROLLMENT IN VOCATIONAL AGRICULTURE CLASSES IN SECONDARY SCHOOLS

Situations Regarding Enrollment	Schools	
	Number	Per Cent
A. Enrollment in vocational agriculture is required of all boys who enter high school	9	10.5
B. Any pupil may elect vocational agriculture if he desires	34	40.0
C. The vocational agriculture course is an alternative to some other course in the high school	2	2.4
D. Enrollment in vocational agriculture is limited to those preparing themselves for farming and having facilities to do adequate supervised practice work including projects	34	40.0
E. The vocational agriculture classes are used to accommodate those pupils who make low grades or who fail in the academic school subjects	1	1.2
F. Any pupil who has facilities to carry a project may elect the course	3	3.5
G. Limited to boys interested in agriculture and having facilities to carry a project	2	2.4
Total	85	100.0

vocational agriculture is given to a selected group of pupils who need it, want it, and are able to profit by it." Forty-seven of the fifty-one persons who responded to the questionnaire on factors characterizing an effective training program in all-day vocational agriculture classes in secondary schools, checked the above-mentioned factor. Thus these educational experts also confirmed the idea that vocational agriculture courses in secondary schools should be confined to a selected group of pupils who need and want the instruction and training, and who can profit by it.

To make a vocational training course required of all boys who enter the high school courses involves a serious waste of time, energy, and money; not only that, it is also contrary to the whole philosophy underlying vocational education, as discussed in Part I of this study.

As shown in Table 7 on page 43, many of the boys in vocational agriculture who are taking the work for some purpose other than that of preparation for farming, and who are in doubt as to the purpose for which they are taking the work, would not be in such classes if adequate restrictions were placed upon the enrollment.

A study of state plans convinces the writer that a more highly selected group of pupils would be found to-day in vocational agriculture if more attention and definite consideration were given in state plans to the kind of pupils that should be enrolled in such classes. There is need in many state plans for clearer and more forceful statements regarding the purpose of instruction in vocational agriculture in all-day schools and classes, and regarding the qualifications of those who should not be admitted into such classes.

Agriculture Subjects Pursued by Boys Enrolled in Vocational Agriculture Classes in Secondary Schools

A great variety of vocational agriculture subjects are pursued by boys in vocational agriculture classes in secondary schools. This point is illustrated in Table 16. Farming differs in different states, and also in different sections of many states. There are, consequently, many different kinds of farming occupations; and it therefore naturally follows that there should be many different kinds of vocational agriculture training courses offered in the secondary schools scattered throughout the nation. Furthermore, in many communities not all the pupils enrolled in the vocational agriculture course desire to prepare themselves for exactly the same kind of farming occupation. This implies that schools must make provision for both group and individual instruction. This particular point is further discussed in Part VIII of this study.

A number of the factors mentioned in Table 2 on page 16 function vitally in this particular connection. These are:

> "Factor 4. The teaching content (subject matter) of the all-day vocational agriculture course is such as specifically functions in the farming occupation for which the pupils are being vocationally trained."
>
> "Factor 5. The supervised training experiences (projects and supplementary farm practice work) in which the pupils enrolled in the all-day vocational agriculture classes engage are such as specifically function in the farming occupation for which the pupils are being vocationally trained."
>
> "Factor 6. The basis of the teaching content of the vocational agriculture courses offered in secondary schools is ob-

TABLE 16

VOCATIONAL SUBJECTS PURSUED BY BOYS ENROLLED IN ALL-DAY VOCATIONAL AGRICULTURE CLASSES IN SECONDARY SCHOOLS

As Reported by Eighty-Eight Teachers of Vocational Agriculture

Vocational Subjects	Number of Schools Offering Subjects in				Total
	First Year	Second Year	Third Year	Fourth Year	
Farm shop work or farm mechanics	29	32	21	9	91
Vocational agriculture, I, II, III, or IV	32	31	18	9	90
Crop production or crops and soils	10	31	1	0	42
Livestock production or animal husbandry	24	12	6	0	42
Farm enterprises	6	9	6	4	25
Farm management	1	0	11	10	22
Agricultural economics	0	1	4	5	10
Fruit production or horticulture	0	4	5	0	9
Agricultural engineering	0	0	4	5	9
Marketing	0	0	2	5	7
Soils	0	0	3	2	5
Poultry	4	0	1	0	5
Dairying	0	2	1	1	4
Gardening	3	0	0	1	4
Special enterprises	0	0	1	1	2
Advanced farm crops	0	0	1	1	2
Advanced animal husbandry	0	0	0	2	2
General agriculture	2	0	0	0	2
Animal and plant life	2	0	0	0	2
Agricultural biology	1	0	0	0	1
Miscellaneous subjects	1	0	0	1	2

tained through the process of job analysis from masters in the occupation (successful farmers in the community), and not from theorists."

"Factor 7. The vocational instruction and training is given on farm enterprises and farm jobs occurring in the farming occupation for which the training is offered, and not on exercises and pseudo jobs."

These four factors, all of which were characterized, as mentioned in Part III, as factors of an effective training program in vocational agriculture, necessitate that the vocational agriculture subjects offered in secondary schools vary.

Summary and Conclusions

The major questions considered in this section were: (1) the length of the all-day vocational agriculture course; (2) school time devoted to vocational agriculture; (3) conditions governing admission into all-day vocational agriculture classes; and (4) agricultural subjects in the vocational courses. These are all administrative problems, governed very largely by the specifications in State Plans for Vocational Education. As a result of a study of twenty-seven State Plans for Vocational Education the writer is of the opinion that there is a need in most of these state plans for more complete, more exact, and more emphatic statements regarding the questions discussed in this section.

PART VIII

PROCEDURES AND METHODS USED IN ALL-DAY VOCATIONAL AGRICULTURE CLASSES IN SECONDARY SCHOOLS IN TRAINING BOYS FOR FARMING

In order to get first-hand facts from teachers of vocational agriculture in secondary schools relative to the procedures and methods they are using in training boys for farming in all-day vocational agriculture classes, five copies of a questionnaire were sent to each of the forty-eight state supervisors of vocational agriculture, with the request that they gather the facts called for on the questionnaire from the teachers of vocational agriculture in the first five schools they visited on their next trip in the field.

Returns were secured on 153 questionnaires from thirty-four states, thus furnishing information from 153 departments of vocational agriculture in secondary schools, upon which this part of the study is based.

The facts gathered in these questionnaires pertained to the following eight points relative to the procedures and methods used in agriculture classes:

1. The dominant objective of vocational education in agriculture in all-day classes, as held by the teachers.
2. The dominant basis of course content.
3. The dominant larger units of instruction.
4. The dominant type of lesson or teaching unit.
5. The dominant method of teaching used in the classroom.
6. The use made of job analysis.
7. The point of view held by vocational agriculture teachers in regard to the purpose and value of the supervised home project work.
8. The extent to which regular school time is set aside for individual project job study and planning.

The Teachers' Points of View Regarding the Objectives of Vocational Agriculture in All-Day Classes

The data in Table 17 show that 72.5 per cent of the 146 teachers of vocational agriculture reported upon in the questionnaire sent to state supervisors of vocational agriculture, stated that the dominant objective characterizing the purpose of the all-day vocational agriculture instruction in schools was: "To train prospective farmers for proficiency in the farming occupation they are planning to follow."

TABLE 17

DISTRIBUTION OF FREQUENCIES SHOWING POINT OF VIEW HELD BY TEACHERS OF VOCATIONAL AGRICULTURE IN REGARD TO THE DOMINANT OBJECTIVE CHARACTERIZING THE PURPOSE OF VOCATIONAL AGRICULTURE IN ALL-DAY CLASSES IN THEIR SCHOOLS

Point of View	Teachers	
	Number	Per Cent
A. To impart general information about farming	4	2.7
B. To give instruction concerning the scientific principles underlying agriculture	9	6.2
C. To fit boys for useful farm employment	23	15.8
D. To train prospective farmers for proficiency in the farming occupation they are planning to follow	106	72.5
E. To create interest in farming	3	2.1
F. To develop attitudes and habits that will fit for useful farm citizenship	1	.7
Total	146	100.0

The results in Table 17 clearly indicate that the great majority of the group of teachers considered in this study held a point of view in regard to the purpose of the work which is in harmony with the underlying philosophy of vocational education; with the aims of the Smith-Hughes Act as discussed in Part I of this study; and with the opinion of 90 per cent of the fifty-one persons who returned the questionnaire on factors characterizing an effective vocational agriculture training program, as given in Factor 1 in Table 2 on page 16.

Bases of Course Content

Table 18 shows that teachers of vocational agriculture use varying bases of course or subject content in the vocational agriculture classes they teach, the dominant basis being "The important farm enterprises of the community with special application to the supervised farm practice work of the pupils."

TABLE 18

DISTRIBUTION OF FREQUENCIES SHOWING THE UNDERLYING BASIS OF COURSE CONTENT IN ALL-DAY VOCATIONAL AGRICULTURE SUBJECTS IN SECONDARY SCHOOLS

Bases of Course Content	Number of Teachers Using Basis	Per Cent
A. Textbooks, corresponding to the subjects (Animal Husbandry, Crops, Soils, etc.)	8	5.4
B. The important farm enterprises conducted by the farmers in the community	20	13.6
C. The important farm problems of the farmers in the community	16	10.9
D. The specific farm enterprises which will be of major importance in the type of farming in which the pupils plan to engage	16	10.9
E. The important farm enterprises of the community with special application to the supervised farm practice work of the pupils	59	40.1
F. The supervised farm practice work (home projects and supplementary farm practice) and the judging activities in which the pupils engage	26	17.7
G. Agricultural bulletins	2	1.4
Total	147	100.0

The theory underlying this basis of course content is that farm boys in a community will enter the type of farming prevailing in the community, and that the supervised project work they engage in will be in the important farm enterprises of the community; thus the procedure would be in accord with Factors 4 and 6 as given in Table 2 on page 16. These factors are as follows:

"Factor 4. The teaching content (subject matter) of the all-day vocational agriculture course is such as specifically functions in the farming occupation for which the pupils are being vocationally trained."

"Factor 5. The supervised training experiences (projects and supplementary farm practice work) in which the pupils enrolled in the all-day vocational agriculture classes engage are such as specifically function in the farming occupation for which the pupils are being vocationally trained."

Only a small per cent of the teachers of vocational agriculture, considered in this study, base their instruction on textbooks; apparently teachers of vocational agriculture use textbooks only as a supplementary aid in instruction.

Some teachers of vocational agriculture are basing their instruction upon "The specific farm enterprises which will be of major importance in the type of farming in which the pupils plan to engage." It is this basis which characterizes the most effective program in vocational agriculture—assuming, of course, that each pupil will also engage in supervised home project work in these specific farm enterprises.

The Dominant Larger Units of Instruction

According to the data in Table 19, farm enterprises, such as pork production, potato production, corn production, are made, by the great majority of teachers of vocational agriculture reported upon, the larger units of instruction in the agriculture subjects. This procedure is in accord with the factor characterizing an effective training program in agriculture in all-day classes which says, as given in Table 2 on pages 16-17, that "The vocational instruction and training is given on farm enterprises and farm jobs occurring in the farming occupation for which the training is offered, and not on exercises or pseudo jobs."

If in item D, in Table 19, Boys' Projects, it is assumed that these projects are real farm enterprises, there is in reality no

TABLE 19

DISTRIBUTION OF FREQUENCIES SHOWING THE DOMINANT LARGER UNITS OF INSTRUCTION CHARACTERIZING THE ORGANIZATION OF SUBJECT MATTER IN ALL-DAY VOCATIONAL AGRICULTURE SUBJECTS

Larger Units of Instruction	Number of Teachers Using Unit	Per Cent
A. Farm enterprises (Such as poultry production, corn production, etc.)	90	59.6
B. Subjects or topics (Such as Feeds and Feeding, Breeds of Livestock, Soil Fertility, etc.)	22	14.6
C. Chapters in a textbook	0	0
D. Boys' projects	34	22.5
E. Major farm problems of community	5	3.3
Total	151	100.0

difference between items A and D as given in the table; the two become synonymous.

Farm enterprises and boys' projects involving the production and disposal of agricultural products are better larger units of instruction in vocational agriculture classes than are subjects, topics, or chapters in textbooks.

Lesson or Teaching Units in Vocational Agriculture Classes

With few exceptions the teachers of vocational agriculture make a farm job a teaching unit, as is shown in Table 20. This procedure is also in accordance with a factor characterizing an

TABLE 20

Distribution of Frequencies Showing the Dominant Lesson or Teaching Unit Characterizing the Lessons in the All-Day Vocational Agriculture Classes

Lesson or Teaching Units	Number of Teachers Using Lesson	Per Cent
A. A farm job or a farm problem	127	83.5
B. An exercise or a pseudo job	5	3.3
C. A topic	4	2.6
D. A number of pages in a textbook or bulletin	1	.7
E. An assigned series of questions to be studied and discussed or recited upon in class	13	8.5
F. A case	1	.7
G. Variation of units—jobs, problems, and cases	1	.7
Total	152	100.0

effective training program in vocational agriculture, as mentioned in Table 2, pages 16-17. This is: "The vocational instruction and training is given on farm enterprises and farm jobs occurring in the farming occupation for which the training is offered, and not on exercises and pseudo jobs." The farm job, whether it be operative or managerial, is a natural unit of work occurring in a farm enterprise or in the management of a farm business, and natural units are better teaching units than are artificial ones.

Methods of Teaching Used in Vocational Agriculture Classes

Class discussion based on a farm job or a farm problem characterizes the dominant use made of the class period in all-day vocational agriculture classes, as is shown in Table 21. The returns on the questionnaire, as recorded in Table 21, show that in only three of the 151 departments of vocational agriculture as reported on the questionnaire, does the recitation based on an assignment in a text or a bulletin, or the lecture method dominate in the teaching. The table further shows that in only five out of 151 cases reported does the method of answering a series of assigned questions predominate in the classroom.

TABLE 21

DISTRIBUTION OF FREQUENCIES SHOWING THE DOMINANT METHOD CHARACTERIZING THE USE OF THE CLASS PERIOD IN VOCATIONAL AGRICULTURE

Method Used in the Class Period	Number of Teachers Using Method	Per Cent
A. Recitation based on an assignment in a book or bulletin	2	1.3
B. Discussion based on an assignment in a book or bulletin	15	9.9
C. Discussion based on a farm job or a farm problem	127	84.1
D. Answering a series of assigned questions	5	3.3
E. Oral presentation by the teacher	1	.7
F. Variation of methods to meet particular unit of instruction	1	.7
Total	151	100.0

It is a generally accepted fact among educators in the field of vocational agriculture that discussion based on an important farm job or on a farm problem, occurring in an important farm enterprise found in the community type of farming, this enterprise being one in which the boys in the vocational agriculture classes have supervised projects, is a method characterizing effective use of the class period as far as class or group instruction is concerned. As will be discussed later in this part of the study, individual instruction based on important operative jobs and managerial problems arising in the boys' supervised practice work also characterizes highly effective use of the class period in vocational agriculture in secondary schools.

If in the use of the class period, as discussed in the preceding paragraph, the purpose of class instruction is essentially concerned with the development of the important manipulative habits (skills) and thinking habits required for competency in the farming occupation for which the instruction and training is being offered (Factor 8 on page 17), then very efficient use is being made of the class period in vocational agriculture classes.

The Use Made of Job Analysis by Teachers of Vocational Agriculture

The particular points investigated to throw light on the question of the use of job analysis by teachers of vocational agriculture were as follows:

1. Has the teacher made a farm survey (occupational survey) of his community?

2. Does the teacher use in course construction the classified data obtained from such a survey?

3. Does the teacher in his teaching differentiate between operative and managerial jobs?

4. To what extent does the teacher analyze the operative job he teaches, as in Bulletins 103 and 118 by the Federal Board for Vocational Education?

5. To what extent does the teacher, with the pupils in his classes, develop analyses of managerial jobs taken up for instruction, as outlined in Bulletins 101, 103, and 105 by the Federal Board for Vocational Education?

6. To what extent do the pupils in the vocational agriculture classes, in their individual project job study and planning, analyze the important operative and managerial jobs not taken up in class, as outlined in Bulletins 101, 103, 105, and 118 by the Federal Board for Vocational Education?

Table 22 shows the replies received to the questions stated above as recorded on the questionnaires. Data in the table would indicate that the great majority of vocational agriculture instructors reported upon make use of job analysis in their teaching as a means of determining purposeful and functioning teaching content.

The great majority of the persons asked to check the questionnaire pertaining to the factors characterizing an effective training program in vocational agriculture in all-day classes checked Factor 6, as shown on page 16. This factor is: "The basis of the teaching content of the vocational agriculture courses offered in secondary schools is obtained through the process of job analysis from masters in the occupation, and not from theorists." The replies from the teachers in regard to the use they

TABLE 22

DISTRIBUTION OF FREQUENCIES SHOWING THE USE MADE OF JOB ANALYSIS BY TEACHERS OF VOCATIONAL AGRICULTURE IN TEACHING VOCATIONAL AGRICULTURE SUBJECTS TO ALL-DAY PUPILS

Points Considered	Teachers Replying				Total	Per Cent
A. The teacher in his teaching differentiates between operative and managerial jobs	Yes 118	To some extent 2	No 33		153	Yes 77.1
B. The teacher analyzes the operative jobs he teaches, as in Buls. 103 and 118 of the Federal Board for Vocational Education	Always 42	Frequently 77	Occasionally 30	Never 4	153	Always and frequently 77.1
C. The teacher with his pupils develops analyses of managerial jobs, as in Buls. 101, 103, and 105 of the Federal Board for Vocational Education	Always 28	Frequently 90	Occasionally 31	Never 4	153	Always and frequently 77.1
D. The pupils in their individual project job study and planning analyze the important operative and managerial jobs not taken up in class	Always 36	Frequently 48	Occasionally 55	Never 13	152	Always and frequently 55.3
E. The teacher makes a farm survey of the community	Yes 118		No 34		152	Yes 77.6
F. The teacher uses in course construction the classified data from his survey	Yes 112		No 40		152	Yes 73.7

make of job analysis would indicate that the majority of the teachers conform to the practice as formulated in the statement of the above factor, and as suggested in the various bulletins issued by the Federal Board for Vocational Education which pertain to the teaching of vocational agriculture in secondary schools.

The Point of View Held by Teachers of Vocational Agriculture Regarding the Purpose and the Value of the Supervised Home Project Work

As shown in Table 23, one hundred and twelve of the teachers of vocational agriculture, or 76.1 per cent, reported that they considered the dominant purpose and value of the supervised home project work conducted by the boys in their vocational agriculture classes to be: "The important training device by means of which the pupils will acquire experience in the management of farm enterprises and skill in the execution of the various jobs involved in such enterprises."

Only five teachers of vocational agriculture considered the supervised project work as something to be done because it is required. These facts clearly indicate that teachers of vocational agriculture have the right point of view regarding the value and the purpose of the supervised home project work.

TABLE 23

DISTRIBUTION OF FREQUENCIES SHOWING THE DOMINANT POINT OF VIEW HELD BY TEACHERS OF VOCATIONAL AGRICULTURE REGARDING THE PURPOSE AND VALUE OF THE SUPERVISED HOME PROJECT WORK CONDUCTED BY THE BOYS IN THE VOCATIONAL AGRICULTURE CLASSES

Point of View Regarding the Purpose and Value of the Supervised Home Project Work	Number of Teachers Holding Point of View	Per Cent
A. As something to be done because it is required	5	3.4
B. As an opportunity for the pupils to earn money while going to school	4	2.8
C. As an opportunity for the pupils to obtain some practical farm experience	16	10.9
D. As an opportunity for the pupils to apply the theoretical facts and principles discussed in the classroom	10	6.8
E. As the important training device by means of which the pupils will acquire experience in the management of farm enterprises and skill in the execution of the various jobs involved in such enterprises	112	76.1
Total	147	100.0

Other dominant purposes of this work as reported by teachers of vocational agriculture, as indicated in Table 23, were:

1. An opportunity for pupils to earn money while going to school.

2. An opportunity for pupils to obtain some practical farm experience.

3. An opportunity for the pupils to apply the theoretical facts and principles discussed in the classrooms.

No doubt all these purposes have a place and play a part in supervised farm project work; they should, however, be regarded as subordinate to the outstandingly important function of such work, and were so regarded by the teachers reporting them. The following quotation clearly emphasizes the real outstanding purpose of the supervised farm practice work:

> "All effective vocational training requires training in both theory and practice—in doing and in thinking about doing. Practice in doing anything is needed to explain and fix theory, while theory is necessary to guide and improve skill. Both are necessary to real job intelligence. The more intimately and closely theory (knowledge) and practice (skill) can be related, the more effective will be the training in each, and the more resourceful will be the job intelligence developed in the student and worker. If you want to become a good golf player, you must practice golf while you think golf, read about golf and take lessons in the theory and practice of golf. Similarly, if you want to train a farm boy to become a successful dairyman, you must have him take care of cows as they should be handled, *while at the same time* you teach him the functioning facts he needs to know and use in his work."[1]

Regular School Time Set Aside for Individual Project Job Study and Planning

Table 24 shows that the great majority of teachers of vocational agriculture frequently (perhaps once or twice a week) set aside regular school time for individual project job study and

TABLE 24

DISTRIBUTION OF FREQUENCIES SHOWING THE EXTENT TO WHICH TEACHERS OF VOCATIONAL AGRICULTURE SET ASIDE REGULAR SCHOOL TIME FOR INDIVIDUAL PROJECT JOB STUDY AND PLANNING

Time Devoted to Individual Project Job Study and Planning	Teachers	
	Number	Per Cent
Daily	12	7.9
Frequently	100	66.2
Occasionally	38	25.2
Never	1	.7
Total	151	100.0

planning in connection with the supervised home project work the pupils are conducting. One hundred and twelve of the one

[1] Prosser, Charles A. and Allen, Charles R., *Vocational Education in a Democracy*, pp. 275-276.

hundred and fifty-one teachers reported that they set aside time for such work daily or frequently.

That individual project job study and planning—that is, the real project method of teaching—can be easily and most effectively used in vocational classes in agriculture, and that much school time should be given over to this type of instruction is recognized by many prominent educators in the field of vocational education. Not only is this method highly advocated for vocational schools but it is highly recommended in other courses in schools, as the following quotation clearly indicates:

> "The curriculum should be conceived, therefore, in terms of a succession of experiences and enterprises having a maximum of lifelikeness for the learner. The materials of instruction should be selected and organized with a view of giving the learner that development most helpful in meeting and controlling life situations. Learning takes place most effectively and economically in the matrix of a situation which grips the learner, which is to him vital—worth while. Traits learned in a natural, or lifelike setting, give promise of emerging definitely in appropriate conduct. It is the task of the teacher and the curriculum-maker, therefore, to select and to organize materials which will give the learner that development most helpful in meeting and controlling life situations. The method by which the learner works out these experiences, enterprises, or exercises should be such as calls for maximal self-direction, assumption of responsibility, of exercise of choice in terms of life values."[2]

Summary and Conclusions

The data pertaining to procedures and methods used by teachers of all-day vocational agriculture classes in training boys for farming, as shown in Tables 17 to 24, clearly indicate that the great majority of these teachers are following procedures and are using methods which are in harmony with the best prevailing practices suggested for this type of work. The procedures and methods used by the great majority of these teachers are also in accord with the factors characterizing an effective training program in vocational agriculture as far as these pertain to procedures and methods of instruction. In this particular aspect of work the great majority of the teachers of vocational agriculture reported upon, rank high.

[2] National Society for the Study of Education, *The Twenty-Sixth Yearbook*, Part II, pp. 18 and 19.

PART IX

THE GENERAL EDUCATION OF BOYS ENROLLED IN ALL-DAY VOCATIONAL AGRICULTURE COURSES IN SECONDARY SCHOOLS

Three questions regarding the nature of the general (non-vocational) education of boys enrolled in all-day vocational agriculture courses in secondary schools were investigated. These were:

1. What is the dominant number of non-vocational subjects taken each year by the boys enrolled in all-day vocational agriculture classes?
2. What is the dominant total number of hours per week of regular school time which boys enrolled in all-day vocational agriculture classes devote to non-vocational subjects?
3. What are the most common non-vocational subjects taken each year by boys enrolled in the all-day vocational agriculture classes?

Information on these three questions was secured from eighty-eight teachers of vocational agriculture on the same questionnaire which was used to secure the facts discussed in Parts VI, VII, and X of this study.

The Dominant Number of Non-Vocational Subjects Taken Each Year by Boys Enrolled in All-Day Vocational Agriculture Classes

As shown in Table 25, in the majority of the departments of vocational agriculture in secondary schools which were studied, the dominant number of non-vocational subjects taken by boys enrolled in the all-day vocational agriculture classes was three. This fact implies that, together with the vocational work, the boys enrolled in all-day vocational agriculture classes, in the majority of the schools studied, take three non-vocational subjects. The nature of these non-vocational subjects is discussed on pages 73 to 76.

TABLE 25

DISTRIBUTION OF THE DOMINANT NUMBER OF NON-VOCATIONAL SUBJECTS TAKEN EACH YEAR BY BOYS ENROLLED IN VOCATIONAL AGRICULTURE CLASSES IN SECONDARY SCHOOLS

As Reported by Eighty-Eight Teachers of Vocational Agriculture *

School Year	Number of Non-Vocational Subjects *Classified by Schools*				Total
	One	Two	Three	Four	
First	0	26	55	5	86
Second	0	17	56	6	79
Third	0	9	41	6	56
Fourth	0	1	26	5	32
Total	0	53	178	22	253
Per cent	0	20.9	70.4	8.7	100.0

* In some schools the number of non-vocational subjects taken by boys enrolled in the vocational agriculture course varies in the different school years.

Hours per Week Devoted to Non-Vocational Subjects by Boys Enrolled in All-Day Vocational Agriculture Classes

According to Table 26, in the majority of secondary schools studied, boys enrolled in vocational agriculture classes devote from 10.1 to 15 hours of regular school time per week to non-vocational subjects. According to Table 13, on page 53, they devote from 5 to 7.5 hours per week to vocational work in agriculture. In other words, in the majority of the departments of vocational agriculture in secondary schools which were studied the boys enrolled in all-day vocational agriculture classes devote about twice as much regular school time to general education as they do to vocational education. This, however, does not take into consideration the time each pupil devotes to supervised project work, which he does outside of regular school hours and which in all states, as specified in State Plans for Vocational Education (see page 33), must be continued for a period of at least six months per year.

The Smith-Hughes Act does not specify the amount of school time in the all-day preparatory vocational agriculture course which shall be devoted to vocational work. In the case of vocational trade, industrial and home economics education the Act does specify the amount of regular school time to be devoted

TABLE 26

DISTRIBUTION OF NUMBER OF HOURS OF REGULAR SCHOOL TIME, PER WEEK, DEVOTED TO NON-VOCATIONAL SUBJECTS BY BOYS ENROLLED IN ALL-DAY VOCATIONAL AGRICULTURE CLASSES IN SECONDARY SCHOOLS

As Reported by Eighty-Eight Teachers of Vocational Agriculture *

School Year	Number of Hours Per Week Devoted to Non-Vocational Subjects *Classified by Schools*				Total
	5–10	10.1–15	15.1–20	20.1 and up	
First	22	45	7	7	81
Second	13	43	8	7	71
Third	7	36	5	5	53
Fourth	2	21	3	5	31
Total	44	145	23	24	236
Per cent	18.6	61.4	9.8	10.2	100.0

* In some schools the time varies in the different school years.

to vocational work, as is shown in the following quotation taken from Section 12 of the Act:

> "That such schools or classes (trade, industrial, and home economics) giving instruction to persons who have not entered upon employment shall require that at least half of the time of such instruction be given to practical work on a useful or productive basis."

Undoubtedly, because of the absence of any such statement as to time requirement in the Smith-Hughes Act in regard to agricultural education, double the amount of regular time in school is given to non-vocational education in the agriculture course in the majority of schools investigated in this study. This statement applies, of course, only to those school years during which the pupils are taking vocational work; in the other years they devote all their school time to general education.

General Education Subjects Studied by Boys Enrolled in Vocational Agriculture Classes

The different kinds of general education subjects taken by boys enrolled in all-day vocational agriculture courses in secondary schools, as reported by eighty-eight teachers of vocational agriculture, are recorded in Table 27. These data were secured from teachers of vocational agriculture on the same questionnaire which was used to secure the information discussed in Parts VI, VII, and X of this study. Owing to the indefiniteness of the replies, specific names of school subjects were not given in many instances.

Table 27 shows the order of the most common general education subjects included in the courses pursued by boys enrolled

TABLE 27

General Educational Subjects (Non-Vocational) Pursued by Boys Enrolled in All-Day Vocational Agriculture Classes in Secondary Schools

As Reported by Eighty-Eight Teachers of Vocational Agriculture

Subject	Number of Teachers Reporting Subjects				Total
	First Year	Second Year	Third Year	Fourth Year	
English	79	69	63	37	248
Algebra	66	13	1	0	80
Geometry	2	47	17	1	67
History	9	16	11	8	44
American history	1	5	25	12	43
General science	29	4	3	1	37
Physics	0	1	11	23	35
Civics	10	3	8	9	30
Chemistry	0	0	18	8	26
Biology	1	17	5	3	26
Commercial subjects	1	3	9	9	22
Foreign languages	5	7	6	2	20
Sociology	3	0	6	7	16
World history	5	6	0	1	12
Modern history	0	10	1	0	11
Physical and health education	4	4	1	1	10
Ancient history	3	6	0	0	9
Problems of democracy	0	0	2	6	8
Science	1	2	3	2	8
Mathematics	1	2	1	1	5
General mathematics	3	1	0	0	4
European history	0	1	1	0	2
Advanced algebra	0	0	0	1	1
Vocations	0	1	0	0	1
English history	0	1	0	0	1
Miscellaneous	7	4	7	15	33

in vocational agriculture classes, as reported by eighty-eight teachers of vocational agriculture.

The ten most common subjects, arranged in order of importance, are:

1. English
2. Algebra
3. Geometry
4. History
5. American history
6. General science
7. Physics
8. Civics
9. Chemistry
10. Biology

It is obvious at once that this list of subjects, with few exceptions, does not correspond at all closely to the list given on page 23 as designated by the experts in the field of general education. The experts' list is repeated here for the sake of ready comparison:

1. English
2. Physical and health education
3. Social problems or sociology
4. Civics
5. General science
6. General mathematics
7. Biology
8. American history
9. Economics
10. World history

Knowing that experts in the field of general education, and particularly those in the field of secondary education, are in a position to decide what general educational subjects are of most worth to boys enrolled in all-day vocational agriculture courses in secondary schools, the writer, as was mentioned in Part IV, sent a questionnaire to sixty such experts; the opinions of the forty-seven who replied were listed in Table 4 on page 24. As has already been mentioned, as far as the results obtained in this study reveal, the majority of boys enrolled in all-day vocational agriculture classes in secondary schools are not, with few exceptions, pursuing the general educational subjects best suited to effective civic, social, cultural, and health training. This same general idea is revealed in the following quotations:

> "To date, the inclusion of vocational agriculture as a part of the public-school program can be characterized as little more than a 'patch' on the existing school organization. Essentially, what has happened is that foreign-language subjects have been withdrawn from the curriculum for vocational pupils and agricultural subjects inserted. The adherence to the traditional high-school organization and to the dominant aim of preparing persons for college has prevented, in many instances, the development of a curriculum in vocational agriculture designed to fit young men for farming occupations. The substi-

tution of agricultural subjects for foreign languages represents the path of least resistance. Doubtless this condition will not be improved until school officials set up and adhere to aims for secondary education that include content other than that designed primarily for college preparation.

"In many schools, principals and teachers have gained the broader conception of the work of a vocational department. Here honest efforts are being made to select content, in such subjects as history, English, science, and mathematics, that will function most effectively in the life activities of vocational pupils. It is at once apparent that close coöperation is needed between the academic teachers and the teacher of agriculture. If both groups of teachers possess a wholesome attitude toward the problem and an earnest desire to select appropriate and functioning content, much may be accomplished."[1]

"The vocational curriculum for prospective farmers in the country high school differs from the college preparatory curriculum of the city high school chiefly by the substitution of studies in agricultural technology for studies in foreign languages."[2]

"A change in the constants in the curriculum of rural and semi-rural high schools is to be desired. The constants should not be those subjects ordinarily required for college entrance, but rather those subjects possessing greatest values for the citizen of the community in the United States. They should be those subjects, or speaking more accurately, that subject matter of greatest value to all, as shown by an analysis of the demands made upon the individual by the everyday activities of present-day life. On this basis the constant subjects, those required of all pupils of a four-year high school, would be some such group as the following: English, community civics, general science, American history and government, economics and sociology, physical education and health, and extra class activities."[3]

Summary and Conclusions

Data secured from eighty-eight teachers of vocational agriculture in secondary schools show that the great majority of boys enrolled at the present time in all-day vocational agriculture classes study, during each year for which they are enrolled in vocational agriculture classes, three non-vocational subjects in

[1] Stewart, R. M. and Getman, A. K., *Teaching Agricultural Vocations*, p. 56.

[2] Eaton, Theodore H., *Vocational Education in Farming Occupations*, p. 168.

[3] U. S. Bureau of Education, *The Rural High School, Its Organization and Curriculum*. Bulletin 1925, No. 10.

addition to the vocational subject; and that these boys devote about twice as much regular school time to non-vocational work as they do to vocational work, during the school years through which they are pursuing vocational work.

The data also show that the non-vocational subjects pursued by boys enrolled in all-day vocational agriculture classes in secondary schools correspond in practically all respects to typical academic work such as is pursued by non-vocational pupils in the secondary schools.

PART X

THE ADMINISTRATION OF THE ALL-DAY VOCATIONAL AGRICULTURE COURSE IN SECONDARY SCHOOLS

In an attempt to discover how rigid and how elastic is the administration of the all-day vocational agriculture courses in secondary schools, the following questions were investigated:

1. Must the boys who enroll in the all-day vocational agriculture course in secondary schools conform to the same entrance requirements (time of entering in the fall and graduation from eighth grade) as do those who take the college preparatory course?
2. Must the boys who enroll in the all-day vocational agriculture course conform to the same attendance requirements (stay in school until the regular school year closes in May or June) as do those enrolled in the college preparatory course?
3. Must all the pupils enrolled in the all-day vocational agriculture classes in secondary schools take the same kind and amount of class instruction in agriculture?
4. May boys enroll for work in vocational agriculture only, without being required to take also some work in non-vocational subjects?

Facts considered essential to the solution of the above questions were secured from teachers of vocational agriculture on the same questionnaire which was used to secure the information assembled in Parts VI, VII, and IX of this study.

Entrance Requirements of All-Day Vocational Agriculture Courses

Table 28 shows the responses of teachers of vocational agriculture who answered the question: "Must the boys who enroll in vocational agriculture classes conform to the same entrance requirements (time of entering school in the fall and graduation from eighth grade) as do those who take the college preparatory course?" Eighty-six of eighty-seven teachers who answered the question, or 98.9 per cent, replied "Yes" to the above question. This fact shows merely that in practically every instance considered in this study the vocational agriculture course in secondary schools is administered, as far as entrance requirements are concerned, just as is the college preparatory, or the regular high school course. Whether or not this should be so will be discussed later in this section of the study.

Attendance Requirements in All-Day Vocational Agriculture Courses

As is shown in Table 28, approximately 95.4 per cent of the eighty-seven teachers of vocational agriculture classes in secondary schools who answered the question: "Must boys who

TABLE 28

DISTRIBUTION OF FREQUENCIES SHOWING THE RIGIDITY AND ELASTICITY OF THE ADMINISTRATION OF THE ALL-DAY VOCATIONAL AGRICULTURE COURSE IN SECONDARY SCHOOLS

Factors Considered	Number of Teachers Reporting			
	Total	Yes	No	Per Cent Yes
A. Must the boys who enroll in vocational agriculture classes conform to the same entrance requirements (time of entering in the fall and graduation from eighth grade) as do those who take the college preparatory course?	87	86	1	98.9
B. Must the boys who enroll in vocational agriculture classes conform to the same attendance requirements (stay in school until the regular school year closes in May or June) as do those who enroll in the college preparatory course?	87	83	4	95.4
C. Must all the pupils enrolled in vocational agriculture classes take the same kind and amount of class instruction in agriculture?	88	58	30	65.9
D. May boys enroll for work in agriculture only, without being required to take also some work in non-vocational subjects?	86	21	65	23.9

enroll in vocational agriculture classes conform to the same attendance requirements (stay in school until the regular school year closes in May or June) as do those enrolled in the college preparatory course?" answered "Yes"; thus indicating again that as far as attendance requirements of pupils is concerned, in the schools studied the vocational agriculture course is administered as is the college preparatory course. The advisability of this procedure is discussed later in this section of the study.

The Kind and the Amount of Class Instruction Required of All Boys Enrolled in All-Day Vocational Agriculture Classes

In the majority of the eighty-eight departments of vocational agriculture (65.9 per cent), as is shown in Table 28, all boys enrolled in vocational agriculture classes must take the same kind and the same amount of class instruction in agriculture. This fact indicates merely that as far as class or group instruction is concerned, the procedure in the great majority of vocational agriculture classes reported upon by teachers of vocational agriculture does not vary from the procedure in non-vocational classes. This point further illustrates, again, that in the great majority of cases studied the administration of the vocational agriculture course corresponds to that of other courses in the secondary school. This question will be discussed further later in this study.

Enrollment in Both Vocational and Non-Vocational Subjects

According to the data in Table 28, in 76.1 per cent of the schools reported by teachers of vocational agriculture, boys may not enroll for work in agriculture only, without taking also some work in non-vocational subjects; also, in two other instances, teachers returning the questionnaire replied that occasionally boys may, by special arrangement, enroll for work only in agriculture. These facts merely indicate, as far as the schools considered in this study are concerned, that boys expecting to study vocational agriculture in the secondary school, must in the great majority of cases pursue a regular course of study including both vocational and non-vocational subjects. Whether or not this should be so will be taken up in the discussion which follows.

The Administration of a Vocational Training Should Be Elastic and Not Rigid

Forty-five of the fifty-one persons who returned the questionnaire on factors characterizing an effective training program in vocational education in agriculture for all-day pupils in secondary schools checked Factor 3, as is shown in Table 2 on page 16, thus indicating that in their opinion this factor characterizes such a program. This factor reads as follows: "The administration of the all-day vocational agriculture course is elastic and not rigid or highly standardized." The above factor implies:

1. That entrance requirements should be elastic, and that any pupil who needs, wants, and can profit by vocational education in agriculture should be admitted into such classes at times when he can best afford to enter and when he will profit by the instruction.

2. That attendance requirements be elastic, thus permitting, if necessary, some boys to leave school before it closes in May or June.

3. That all boys in all-day vocational agriculture classes should not be required to take exactly the same kind and the same amount of class instruction; opportunities should assume that they are not all preparing for exactly the same kind of farming occupation.

4. That boys should be allowed to enroll only for vocational work if circumstances are such that they have only a limited time for schooling and if they desire most the vocational work.

When conditions such as those just enumerated prevail, it should be clearly understood that boys not complying with all the regular school requirements would be regarded not as candidates for a regular four-year high school diploma, but rather as special pupils attending school for a special purpose and for a limited amount of time.

The administration of such an elastic program in vocational agriculture is possible if much of the school period devoted to vocational work is given over to individual self-education (individual instruction), based on the supervised farm project work of the boys enrolled in the agriculture classes. The writer believes that a farm boy sixteen years of age who is interested in dairy farming, but who is unable to go to school except between October and April, and who has excellent opportunities of carrying on a good dairy project, should be allowed to enroll in the vocational class in animal husbandry offered in the local high school to receive instruction in farm dairying. He also believes that a farm boy seventeen years of age, intensely interested in farming but not able to take the regular four-year vocational agriculture course in the local high school, should be allowed to devote all his school time to vocational agriculture.

Summary and Conclusions

In practically all respects the all-day vocational agriculture course in secondary schools, as reported upon by teachers of vocational agriculture, is administered just as is the college preparatory course in secondary schools. Pupils expecting to prepare themselves for farming must conform to the same entrance and attendance requirements as must any other pupil in the school. Also, all boys must take exactly the same kind of class instruction. These practices are contrary to those factors in an effective training program in vocational agriculture which say: "The administration of the all-day vocational agriculture course is elastic and not rigid or highly standardized"; "The vocational training helps the pupils to capitalize their vocational interests, aptitudes, and abilities to the highest degree"; and "The instructional needs of the pupils are met at the time when these pupils most require help, and individual instruction is given whenever necessary to the progress of any pupil." The present-day administration of vocational classes in agriculture should therefore be made less rigid and far more elastic.

PART XI

GENERAL CONCLUSIONS

The Smith-Hughes Act was passed for a very specific purpose, that of promoting and encouraging vocational education of less than college grade for useful employment in the home, in the trades, in the industries, and in farming. To give young people, looking forward to farming as an occupation, an opportunity to prepare themselves for useful farm employment, in schools and classes under public control and supervision, is the purpose of vocational education in agriculture in all-day classes in secondary schools as promoted and encouraged under the provisions of the Smith-Hughes Act. There is only one measure of the effectiveness of a preparatory vocational training course, and that is the degree to which it prepares those taking the vocational instruction and training for the occupation for which it is given.

Since young people enrolled in all-day federally-aided vocational agriculture courses in secondary schools are taking both vocational training for farming and general educational training for effective citizenship, it follows that the schooling and training these young people are receiving must, in order to be effective, contribute specifically and vitally to both of these purposes.

Factors Characterizing an Efficient Vocational Training Program in Agriculture in All-Day Classes

In the opinion of 76 per cent, or more, of fifty-one persons who may well be regarded as a group of experts in the field of vocational education, the thirteen factors mentioned in Table 2 on pages 16 and 17 are characteristic of effective vocational instruction and training in agriculture of less than college grade for all-day pupils in vocational agriculture courses in secondary schools. Individually, various members of this group of fifty-one experts are also of the opinion that certain ones of the fifteen factors listed in Table 3 on page 18 also characterizes an effective training program in all-day vocational agriculture classes.

These factors may and generally do characterize conditions far in advance of the conditions actually prevailing at the present time in all-day vocational agriculture classes in secondary schools; nevertheless they need to be regarded as ideal conditions toward which training programs in all-day vocational agriculture courses in secondary schools need to seek to conform in order to be highly effective.

General Education Subjects of Most Worth to Boys Taking the Vocational Agriculture Course in Secondary Schools

According to the majority opinion of a group of forty-seven outstanding educators in the field of general education, a group that may well be regarded as one of experts in general secondary education, the ten following subjects, arranged in order of importance, are of most worth to boys enrolled in all-day vocational agriculture courses in secondary schools:

1. English
2. Physical and health education
3. Social problems or sociology
4. Civics
5. General science
6. General mathematics
7. Biology
8. American history
9. Economics
10. World history

It is also the opinion of the majority of the forty-seven experts in the field of general education that only the above mentioned subjects (or subjects enumerated in the list) should be required subjects or constants in all-day courses in vocational agriculture in secondary schools.

Unless state plans for vocational education or special circulars supplementing state plans, such as are referred to on pages 41 and 85, specifically state and recommend the general educational subjects of most worth to boys enrolled in vocational agriculture courses, and incorporate only such subjects as are mentioned in the preceding paragraph, the writer believes that little progress will be made by the schools in giving the best training for effective citizenship.

The curriculum on page 37 comes very close to incorporating the ideas and opinions of the forty-seven experts in general education.

State Plans for Vocational Education in Agriculture

With few exceptions, in the twenty-seven copies of State Plans for Vocational Education which were analyzed and studied, the sections of these plans devoted to vocational education in agriculture in all-day schools or classes were very generalized and vague and entirely lacking in specific and direct statements as to what should characterize this type of work in secondary schools in order to obtain the most efficient results. If such plans are to have much value to a state in guiding teachers and school authorities in the administration of a well-defined program in vocational agriculture, it appears that statements in the great majority of these plans in regard to (1) the purpose of vocational education in agriculture, (2) the type of boys for whom the work is intended, (3) the courses of study, (4) the methods of instruction, and (5) the purpose and the value of the supervised farm project work need to be made very much clearer and much more specific.

State plans, with few exceptions, are brief. They do not go into detail on matters vitally affecting the execution of a good training program in vocational education in agriculture. Therefore a special circular, such as is mentioned on page 41, could be effectively used in all states to emphasize and elaborate upon the five points mentioned in the preceding paragraph. Such a circular (1) could state in much more elaborated terms than do the state plans the purpose of vocational education in agriculture—stating this purpose in some such way as it is developed in Part I of this study; (2) it could define the type of boys for whom vocational agriculture is intended much as was mentioned on pages 43 and 46; (3) it could give illustrations of courses of study such as the two excellent curricula on pages 36 and 37; (4) it could describe in detail methods of instruction somewhat in the way illustrated by item 4 on page 39 and by item 5 on page 40; and (5) it could state the purpose and value of the supervised farm practice work somewhat as the purpose and value are described on pages 67 to 69. Moreover, such a circular could weave into the suggestions and discussions on all the five points just enumerated the thirteen factors characterizing an effective vocational training program stated on page 11.

The Type of Boys in All-Day Vocational Agriculture Classes

In this study data were secured on 2,809 boys enrolled in federally-aided vocational agriculture classes in secondary schools. Of this number 78.5 per cent were in the first two-year classes, and 21.5 per cent in the last two year classes. The median age of all these boys was 15.7 years. The median age of the 78.5 per cent of the boys in the first two-year classes was 15.3; and 56.7 per cent of all these 2,809 boys were between the ages of 13 and 16 years of age. These facts suggest a few important questions, namely: (1) Can boys between the ages of 13 and 16 years profit very much by real vocational training? (2) In order to get the most effective results, should the enrollment in vocational agriculture classes in secondary schools be concentrated in the first two years or the last two years of the four-year high school course? (3) Since only 4.6 per cent of the 2,809 boys in this connection were enrolled in the fourth-year class of the secondary school, what happened to the 93.4 per cent? These questions are serious ones; the problems they suggest require further investigation before the questions can be answered authoritatively.

Owing to the fact that high schools have no farm land, livestock, and farm equipment, and that they must depend almost entirely upon home projects in agriculture for supervised farm practice work, it is essential that boys enrolled in all-day vocational agriculture classes should live on farms, or have other adequate farm facilities for project work.

Approximately 31.3 per cent of the boys, considered in this study, who are enrolled in all-day vocational agriculture classes at the present time are not interested in preparing themselves for the occupation of farming, and are in doubt as to why they have enrolled in vocational agriculture classes. This fact is an outstanding weakness in vocational education in agriculture in our secondary schools. It can and should be overcome by strict adherence to definite entrance requirements which conform to the regulations and spirit of the Smith-Hughes Act.

Another flagrant weakness needs to be recognized and eliminated. It is indeed difficult to comprehend how boys can be effectively trained for the occupation of farming by conducting their supervised projects in back yards and on vacant lots in town. The Smith-Hughes Act clearly mentions the fact that "schools shall provide for directed or supervised practice either on a *farm* provided for by the school or other *farm* for at least six months per year."

As was mentioned on page 41, a much more highly selected group of boys would undoubtedly be found in vocational agriculture classes if state plans and special circulars supplementing state plans described more fully the purposes of vocational agriculture and showed the necessity for enrolling only the type of boys for whom the work is intended. The necessity for enrolling the right type of boys in the vocational agriculture course is emphasized and made clear in item 5 on page 39.

The All-Day Vocational Agriculture Course in Secondary Schools

In the great majority of the schools studied which offer instruction in vocational agriculture in all-day classes, 90 minutes per day of regular school time is given over to vocational work for a period varying from two to four school years.

In the great majority of these same schools boys enrolled in all-day vocational agriculture classes take three regular non-vocational subjects together with the work in vocational agriculture. These same boys devote twice as much regular school time to non-vocational or general educational work as they do to vocational work. An important question arises in this connection. Are these boys primarily taking a vocational training course in preparation for farming, or are they taking a general high school course and choosing the vocational work as an elective school subject? Table 7 on page 43 shows that 68.7 per cent of the 2,740 boys considered in this study are taking the work in vocational agriculture for preparation for farming.

The fact has been mentioned, as is shown in Table 7, that 13.7 per cent of 2,740 boys enrolled in vocational agriculture classes, which were considered in this connection, are taking the vocational work for some other purpose than preparing themselves for farming; and that 17.6 per cent of these boys are in doubt as to just why they are taking the work in vocational agriculture. This condition, in the opinion of the writer, indicates another weakness in the present-day administration of the program in question. The writer believes that this weakness is undoubtedly due to the fact, as shown in Table 15 on page 56, that, in about 40 per cent of the schools studied, any pupil may elect vocational agriculture if he so desires, apparently, without compliance with any specific entrance requirements.

On the other hand, in 40 per cent of the schools studied, as shown in Table 15 on page 56, "Enrollment in vocational agriculture is limited to those who are preparing themselves for farming and have facilities to do adequate supervised practice work including projects." It is difficult to see how "The instruction and training in vocational agriculture is given to a selected group of pupils who need, want it, and are able to profit by it" (Factor 3 as given in Table 2 on page 16), unless some such definite restriction as that mentioned in the preceding sentence is made in regard to who should and who should not be permitted to enroll in the vocational agriculture course.

The fact should not be overlooked that directed or supervised practice in agriculture, most of which usually takes the form of home projects, is a required part of the vocational agriculture course. According to the Smith-Hughes Act and the State Plans for Vocational Education this work must be continued for at least six months per year by all pupils enrolled in a federally-aided vocational agriculture class. This work is done outside the regular school time and supervised by the teacher of vocational agriculture, employed on a twelve months' basis.

Suggestions for the improvement of vocational agriculture courses can be found in the details of the two curricula on pages 36 and 37 and in the last quotation on page 39. It seems almost needless to repeat the fact that: (1) the length of the vocational agriculture course; (2) the portion of the school day to devote to vocational work; (3) the admission into vocational agriculture classes; and (4) the subjects in the vocational agriculture course in administrative problems are governed very largely by the stipulations in state plans. Therefore in order to improve conditions in regard to these matters the writer believes that the improvements must first be made in the state plans and in the special circulars supplementing the state plans.

Procedures and Methods Used in Secondary Schools in Training Boys for Farming

The study of present-day methods and procedures used in vocational agriculture classes invites less criticism and censure, however, than does the foregoing phase of the subject under discussion in this study. As recorded in Table 17 on page 61, the dominant point of view held by teachers of vocational agriculture, in 153 schools reported upon, in regard to the purpose of the instruction in vocational agriculture is that that purpose is to train prospective farmers for proficiency in the farming occupation they are planning to follow.

The majority of teachers of vocational agriculture whose departments of vocational agriculture were studied, reported that they make:

1. Important farm enterprises of the community, with special application to the supervised farm practice of the pupils, the dominant basis of the course content in vocational agriculture classes.
2. Farm enterprises the dominant larger units of instruction in vocational agriculture classes.
3. A farm job or a farm problem the dominant teaching unit in vocational agriculture classes.
4. A discussion based on a farm job or on a farm problem the dominant use of the class period.

The majority of the teachers of vocational agriculture whose departments of vocational agriculture were studied believe the dominant purpose and value of the supervised home project work to be "the important training device by means of which the pupils will acquire experience in the management of farm enterprises and skill in the execution of the various jobs involved in such enterprises."

Table 22, page 67, shows that the great majority of the teachers of vocational agriculture reported upon are making use of job analysis in teaching vocational agriculture as suggested in Federal Board for Vocational Education Bulletins Nos. 101, 103, 105, and 118.

The conclusion drawn from the statements just enumerated is therefore in regard to the procedures and methods used by teachers of vocational agriculture in training boys for farming, as the data on Tables 17 and 22 show, that these are in accord with the factors characterizing effective instruction and training in all-day vocational agriculture classes as given in Table 2 on pages 16 and 17.

General Education of Boys Enrolled in Vocational Agriculture Courses in Secondary Schools

In the majority of the schools studied boys enrolled in all-day vocational agriculture classes pursue three non-vocational or general educational subjects along with the school work in vocational agriculture during those school years in which they are taking vocational work, and they devote approximately twice as much regular school time to general educational subjects, or work, as they do to vocational work.

The ten most common general educational subjects pursued by boys enrolled in vocational agriculture classes in secondary schools, as reported by eighty-eight teachers of vocational agriculture, are, in order of popularity:

1. English
2. Algebra
3. Geometry
4. History
5. American history
6. General science
7. Physics
8. Civics
9. Chemistry
10. Biology

As was mentioned on page 75, the list given above does not correspond very closely, either in subjects or in the order of importance, to the list which experts in general education regard as most important for boys enrolled in the vocational agriculture in secondary schools.

Except for the exclusion of the foreign languages, the great majority of boys enrolled in vocational agriculture classes in secondary schools, considered in this study, take practically the same general education subjects as do boys enrolled in the college preparatory course. This fact indicates another weakness of the school instruction offered boys enrolled in all-day vocational agriculture courses in secondary schools. The following quotation corroborates this point of view:

> "If such schools, representative as they seem to be of rural and semi-rural high schools giving four years of high school work, are to meet the varying needs of the pupils enrolled in the average high school, the selection of subjects and the organization of curricula must be carefully planned. The most common practice today, according to all data at hand, is to make the college entrance subjects the core of the work given in the small high school. This has been done in the face of the fact that only about 35 to 40 per cent of those who enter the high school are graduated and that only 17 to 20

per cent enter higher institutions of learning of any description. A phase of this practice of making the college preparatory subjects the constants in the work of the small high school is the placing of these required subjects, such as mathematics, foreign language, ancient and medieval history, mainly in the first two years of the course. The result has been that the subjects required of the many have been particularly suited to the needs of the few, and the subjects available to the few in the last two years have been those of most value in meeting the needs of the many."[1]

The quotations from experts in the field of general education as given on pages 24 to 27 contain many suggestions of value in providing a better balanced and more effective type of general education for boys enrolled in vocational agriculture courses in secondary schools.

As previously mentioned, courses of study in vocational agriculture must conform in general to the stipulation of the State Plans for Vocational Education. The writer is of the opinion that unless the state plans or the special circulars supplementing the state plans give special attention to the general educational subjects which would be of most worth to boys enrolled in vocational agriculture courses, and incorporate these as constants in such courses, the condition referred to in the quotation given above, which is also the condition actually found in practice, will result.

The Administration of Vocational Agriculture Courses in Secondary Schools

In practically every respect, as reported by the great majority of the teachers of vocational agriculture whose departments of vocational agriculture are studied, the vocational agriculture course is administered exactly as is the college preparatory or the regular high school course. In other words, boys wishing to take the all-day vocational agriculture course in secondary schools in order to prepare themselves for the pursuit of farming must (1) be eighth grade graduates, (2) enter school in September when it opens, (3) stay in school until it closes in May or June, (4) take the same kind of class instruction in vocational agriculture as every other prospective farmer, whatever the type of farming he intends to engage in, and (5) carry three non-

[1] U. S. Bureau of Education, *The Rural High School—Its Organization and Curriculum.* Bulletin, 1925, No. 10.

vocational subjects each year along with the vocational agriculture work.

The situation described in the preceding paragraph is most certainly not in accord with Factor 13 as given in Table 2 on page 17. This factor reads as follows: "The administration of the all-day vocational agriculture course is elastic and not rigid or highly standardized."

If it is true that a vocational training course is offered for the sake of those who need it, want it, and are able to profit by it, it becomes apparent that the all-day vocational agriculture course in secondary schools should be so administered that it can render the greatest possible service to the greatest possible number of these young people who want it, need it, and are able to profit by it. To do this it cannot be operated as a highly standardized course and be kept as rigid as is the college preparatory course.

How the vocational agriculture course in secondary schools is to be administered is an administrative problem, influenced wholly in every state by the provision of that State's Plans for Vocational Education. Unless these plans or the special circular supplementing them advocate and highly recommend that the administration of the vocational agriculture course be elastic in order that it may render the greatest possible opportunity to all the youth in a rural community who may be interested in more effectively preparing themselves for some occupation of farming, the schools will continue to administer the vocational agriculture courses just as they have always done, in a manner which is very rigid and highly standardized, conforming in many respects to the academic course. This problem can well be summarized by a statement by J. B. Edmonson, Secretary of the North Central Association, previously quoted on page 28, in which he says: "The fact that certain changes have not been made in secondary schools is traceable in my opinion to the marked conservatism of high school administrators and teachers and to the failure of advocates of new ideas to present convincing arguments."

APPENDIX

A. LISTS OF PERSONS RETURNING QUESTIONNAIRE

I. List of persons who returned the questionnaire on general educational subjects:

Ashbaugh, E. J.
Bagley, W. C.
Bobbitt, Franklin
Bonser, F. G.
Bustow, Wm. H.
Carrothers, George E.
Charters, W. W.
Cook, Katherine M.
Cox, Philip W. L.
De Voss, James C.
Douglass, Harl R.
Dunn, Fannie W.
Eaton, T. H.
Ferris, Emery N.
Freeman, E. N.
Frost, Norman
Gaumintz, W. H.
Hacker, L. W.
Hamin, S. A.
Harap, Henry
Hopkins, L. Thomas
Horn, Ernest
Howard, George
Jackson, G. L.
Jones, Arthur J.
Joygers, R. E.
Kirby, Thomas J.
Kitson, H. D.
Koos, L. V.
Lewis, Melvin S.
Maxwell, C. R.
Monroe, Walter S.
Newlon, Jesse H.
Myers, George E.
Nelson, M. G.
Pittman, M. S.
Reavis, W. C.
Schorling, Raleigh
Snedden, David
Stewart, R. M.
Symonds, Percival M.
Thorndike, E. L.
Toops, Herbert A.
Van Wagenen, M. J.
Webb, L. W.
White, Trentwell M.
Worcester, D. A.

II. List of persons who returned the questionnaire on factors characterizing an effective vocational training program:

Adams, O. D.
Bass, M. R.
Baylor, Adelaide S.
Billington, R. V.
Blackwell, J. D.
Blodgett, Mary
Chapman, Paul
Clements, D. M.
Clements, L. D.
Cotterman, H. F.
Coxen, Jas. R.
Davies, L. R.
Dennis, L. H.
Fife, Ray
Fitzgerald, N. E.
Gardner, H. M.
Gentry, C. B.
Getman, A. K.
Gibson, H. H.
Heald, F. E.
Holloway, Keith L.
Howard, Carl G.
Kane, Michael
Keppel, R. N.
Kerr, William
Klemmedson, L. D.
Magill, Edward C.
Mallary, B. E.
Maltby, Robert D.
McPhee, Julian A.
Miller, C. M.
Miller, Halbert W.
Moore, F. E.
Pearson, James H.

Pollom, Lester B.
Reavis, Geo. W.
Richardson, D. F.
Ross, W. A.
Sampson, H. O.
Sanders, Geo. S.
Sargent, H. O.
Sasman, Lewis M.
Snedden, David
Stewart, W. F.
Stimson, Rufus W.
Sutherland, S. S.
Tieman, Harry A.
Treat, F. M.
White, Marie
Williams, Arthur P.
Williams, C. V.

B. BIBLIOGRAPHY

EATON, THEODORE H. *Education and Vocations.* John Wiley and Sons, Inc., New York, 1926.

FEDERAL BOARD FOR VOCATIONAL EDUCATION:

Agricultural Education—Organization and Administration, Bulletin 13.

Agricultural Project Planning, Bulletin 117.

Analysis of the Management of a Corn-Growing Enterprise, Bulletin 101.

Analysis of the Management of a Cotton-Growing Enterprise, Bulletin 105.

Analysis of the Operative Jobs of a Corn-Growing Enterprise, Bulletin 118.

Methods of Teaching as Applied to Vocational Education in Agriculture, Bulletin 103.

Supervised Practice in Agriculture Including Home Projects, Bulletin 112.

FERRIS, EMERY T. *Secondary Education in Country and Village.* D. Appleton and Company, New York, 1927.

KOOS, LEONARD V. *The American Secondary School.* Ginn and Company, Boston, 1927.

LEE, EDWIN A. and OTHERS. *Objectives and Problems of Vocational Education.* McGraw-Hill Book Company, New York, 1928.

MAYS, ARTHUR B. *An Introduction to Vocational Education.* The Century Company, New York, 1930.

NATIONAL SOCIETY FOR THE STUDY OF EDUCATION. "The Status of Rural Education." *Thirtieth Yearbook,* Part I. Public School Publishing Company, Bloomington, Illinois, 1931.

PROSSER, CHARLES A. and ALLEN, CHARLES R. *Vocational Education in a Democracy.* The Century Company, New York, 1925.

SCHMIDT, G. A. *Efficiency in Vocational Education in Agriculture.* The Century Company, New York, 1928.

SNEDDEN, DAVID. *American High Schools and Vocational Schools in 1960.* Bureau of Publications, Teachers College, Columbia University, New York, 1931.

SNEDDEN, DAVID. *Vocational Education.* The Macmillan Company, New York, 1920.

STEWART, R. M. and GETMAN, A. K. *Teaching Agricultural Vocations.* John Wiley and Sons, Inc., New York, 1930.

U. S. BUREAU OF EDUCATION. *The Rural High School: Its Organization and Curriculum.* Bulletin, 1925, No. 10.

WRIGHT, J. C. and ALLEN, CHARLES R. *Administration of Vocational Education.* John Wiley and Sons, Inc., New York, 1926.

WRIGHT, J. C. and ALLEN, CHARLES R. *Efficiency in Vocational Education.* John Wiley and Sons, Inc., New York, 1929.